ANF	BAL	03/18
1 2 APR 2018 07/20 DS		

This book should be returned/renewed by the latest date shown above. Overdue items incur charges which prevent self-service renewals. Please contact the library.

Wandsworth Libraries
24 hour Renewal Hotline
01159 293388
www.wandsworth.gov.uk

Wandsworth

The Earth series traces the historical significance and cultural history of natural phenomena. Written by experts who are passionate about their subject, titles in the series bring together science, art, literature, mythology, religion and popular culture, exploring and explaining the planet we inhabit in new and exciting ways.

Series editor: Daniel Allen

In the same series

Air Peter Adey

Cave Ralph Crane and Lisa Fletcher

Clouds Richard Hamblyn

Comets P. Andrew Karam

Desert Roslynn D. Haynes

Earthquake Andrew Robinson

Fire Stephen J. Pyne

Flood John Withington

Gold Rebecca Zorach
   and Michael W. Phillips Jr

Islands Stephen A. Royle

Lightning Derek M. Elsom

Meteorite Maria Golia

Moon Edgar Williams

Mountain Veronica della Dora

Silver Lindsay Shen

South Pole Elizabeth Leane

Storm John Withington

Swamp Anthony Wilson

Tsunami Richard Hamblyn

Volcano James Hamilton

Water Veronica Strang

Waterfall Brian J. Hudson

Swamp

Anthony Wilson

REAKTION BOOKS

To Jeanette and Lucas

Published by Reaktion Books Ltd
Unit 32, Waterside
44–48 Wharf Road
London N1 7UX, UK
www.reaktionbooks.co.uk

First published 2018

Printed and bound in China

A catalogue record for this book is available from the British Library

ISBN 978 1 78023 844 9

CONTENTS

Introduction: Terra Incognita

What is a swamp?

Answering this seemingly simple question proves fascinatingly difficult. Swamps have always, by their very nature, resisted easy classification. They vary from continent to continent, from culture to culture – in some places, great, sprawling survivals of the pristine or primordial; in others, pockets of chaos in an increasingly ordered world.

In terms of scientific classification, swamps are easy enough to define. They are essentially wooded wetlands, distinguished from bogs or marshes primarily by the presence and density of trees. This terminology, though, is deceptively simple, in that the term 'wetland' itself was a creation of the 1950s, in part in response to the negative connotations borne by a centuries-old catch-all term for soggy ground – 'swamp'. Generally, people have used terms like 'marsh', 'swamp' and even 'bog' as synonyms.[1] Swamps, marshes, bogs: while they may conjure different images, they have historically been more terms of subjective experience than of precise, objective classification, and of them all, 'swamp' has seen the widest and most varied use.

Before we embrace fully the mercurial and subjective aspects of our opening question, let us consider the scientific and objective ones, as far as they will take us. Swamps are, unquestionably, a variety of wetland, and wetlands are somewhat more amenable to clear definition. 'Wetland' is a broadly inclusive term, encompassing a wide array of ecosystems and environments. They are essentially in-between areas, mixtures of water and earth, land

A Florida swamp.

and liquid. They may be permanently or seasonally saturated, and include swamps, bogs, floodplains, pocosins, marshes, mires and other such places.

The three largest and most significant wetland categories that tend to fall under the general category of 'swamp' include swamps proper, marshes and bogs. Swamps are saturated areas of land distinguished by woody vegetation, primarily trees. There are both fresh- and saltwater swamps, with the former generally located inland and the latter in coastal areas. With the exception of Antarctica, swamps can be found on every continent, and range in size from tiny, isolated pockets of saturated ground to enormous basins and deltas.

Notable freshwater swamps include the swamps of the Fertile Crescent, between the Tigris and Euphrates rivers, an area of incredible biodiversity that saw the birth of human civilization. These swamps provided early humans with abundant fish and wildlife for hunting, as well as navigable waterways to facilitate travel and arable land. The Ma'dan, or Marsh Arabs, have lived among them for thousands of years, raising cattle, creating islands from reeds and coexisting with the enormous wild boars who also inhabit the area.

Many famous freshwater swamps can be found in the southern United States. Perhaps the most well-known among the swamps of the South is the Great Dismal, which spans many thousands of acres in Virginia and North Carolina. This tremendous swamp, with its forests of poplar, pine, cedar, maple and bald cypress arrayed around the central Drummond Lake, also incorporates bogs and marshes. Its mercurial mixture of beauty and menace has been celebrated by painters and poets, memorialized in legend and folk tale; in some ways, it has become the archetypal North American swamp. The Dismal is home to a vast array of wildlife: it is full of mammals like deer, otters, raccoons and foxes, and home to less frequently spotted species of bear and bobcat. Reptiles and amphibians populate the swamp in great numbers and diversity: more than twenty species of snakes, both venomous and benign, live alongside over fifty species of turtles, salamanders, lizards, frogs and alligators. Over two hundred

Musk Rat

Aligator

Bull Frog

Viper

Terebin

Muskeetoe Hawk

Green Lizard

Parekee toe

Tum ble
Tu rd

Rattle Snake

Horn Snake

Bald Eagle

Goffhawk

species of birds make their homes here as well. Swamps like the
Great Dismal are rich in wildlife because of their clean waters,
their abundance of vegetation, and the number and variety of
land-based and aquatic habitats they provide.

Another distinctive United States wetland system lies in
Louisiana, where over 10,000 square miles (35,000 square kilometres) of swamp, bayou and marshland, threatened and shrinking as they are, still constitute over 10 per cent of the United
States' wetlands.

Among the knobby cypress knees that jut from the waters
of the Atchafalaya Basin live black bears, cougars and foxes, as
well as the ratlike nutria. Snakes, including king snakes and
cottonmouths, slither alongside prodigious alligator snapping
turtles; perhaps the most infamous denizens of the Louisiana
swamps, though, are the alligators themselves, great saurians
who lend the swamps much of their fascination and menace.
Birds of the Louisiana wetlands include the osprey, the great
blue heron, the great egret and, of course, the pelican, Louisiana's
state bird. The freshwater swamps provide food, shelter and
sustenance for a wide range of species.

Saltwater swamps, by contrast, are found in coastal areas,
often in tropical climates. In areas regularly inundated by high
tides, trees such as mangroves create networks of roots around
which sand and soil accumulate. Such coastal marshes offer
shelter to many marine creatures, and become spawning areas
for fish and many other species, including shellfish and crabs.
Coastal marshes also attract a wide array of birds, which feed
on the abundant fish and fertilize the soil with their droppings.

Marshes are another type of wetland, distinguished from
swamps by the absence of trees. Instead, marshes are dominated by grasses, reeds and other herbaceous plants. Like swamps,
there are inland and coastal, fresh- and saltwater marshes. Tidal
or coastal marshes have great ecological importance: they slow
coastal erosion and provide habitats and shelter for sea birds and
other wildlife. They also retard the process of saltwater intrusion,
keeping inland fresh waters from being polluted by ocean water
seeping into the water table. Marshes breed abundant insects,

which, in turn, provide food for birds of many kinds, and provide shelter and food for all manner of aquatic life. Inland marshes, usually located near lakes and rivers, vary from seasonally inundated meadows to enormous marsh complexes like the Florida Everglades, also known as the 'River of Grass'. Actually a vast, slow-flowing river, the Everglades provide homes and sustenance for an array of waterfowl as well as amphibians, reptiles and even panthers. Botswana's Okavango Delta, most likely the world's most extensive freshwater marsh, is a sustaining centre for all manner of wildlife. Known as the 'Jewel of the Kalahar', the Okavango Delta provides fresh water for animals ranging from elephants and giraffes to water-dwelling creatures like

American alligator in The Basin, June 2007.

hippopotami and crocodiles. These inundated, grassy areas are rich in biodiversity.

Bogs are a third variety of wetland, distinct in kind from marshes and swamps, but often considered synonymous because of their characteristic mix of land and water. Freshwater wetlands that develop in cooler climates, bogs form over the course of centuries as bodies of water gradually fill up with both live and decaying plants, sphagnum moss and other vegetation. One type of bog is the quaking bog, which forms as moss and vegetation develop mats around half a metre to a metre in thickness on the surface of a pond or other body of water; when stepped on, quaking bogs give and spring back slightly, hence their name. Other types include blanket bogs, which cover large, varied areas of landscape; string bogs, which are studded with islands of solid

Foggy sunrise on the Cocodrie Swamp, October 1991.

earth amid the bog surface; and raised bogs, which take on a domelike shape.

Bogs, while they may not teem with the kind of animal life associated with the Okavango Delta or the Great Dismal Swamp, are actually quite biodiverse. They are home to a wide array of insects, which in turn support a variety of amphibians and birds. Cranberries and other berries grow in some kinds of bog, and while few larger land animals live in them, mammals like moose and otters will often come to bogs to feed.

Okavango Delta (Okavango Grass-land), Botswana, southwest Africa, is one of the Seven Natural Wonders of Africa.

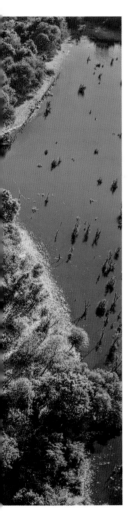

One of the most intriguing features unique to bogs is the formation of peat, which is a kind of intermediate step in the process of decaying moss and vegetation developing into coal. Peat can be harvested and burned for energy. Bogs in Russia, the British Isles and Scandinavia have been used as energy sources for centuries. Peat bogs also sequester carbon, capturing carbon dioxide that would otherwise enter the atmosphere and exacerbate global warming. The destruction of bogs for development or to harvest peat releases carbon, with significant environmental impact. Peatlands are widespread in the northern latitudes: more than 17 per cent of Ireland consists of current or former bog land. Canada is 18 per cent peatland, and England 6 per cent.

If we expand our notion of 'swamp' to include these varied wetlands, the term becomes much more inclusive. While the Great Dismal and the Louisiana swamps more or less embody the term 'swamp' in the Western imagination, the great Vasyugan Mire, a massive system of bogs, forested mires and fens that represents the single largest swamp in the northern hemisphere, lies not in some sultry southern clime, but in Siberia.[2] Representing approximately 2 per cent of the world's peat bogs, the Vasyugan Mire encompasses fens and small forested swamp areas populated with pine, spruce, cedar, fir and birch trees. The average temperature in the Vasyugan Mire is -1.1°C, though summer temperatures can be in the high teens. While fewer species of amphibians and reptiles populate the relatively cold Vasyugan Mire than many other wetland environments, nearly two hundred species of birds and a broad array of mammals, including many small rodents and ranging up in size to lynx and sable, brown bear and elk, make their homes here.[3]

Swamps, great and small, can be found all over the world, from the enormous South American Pantanal, a swamp system that spans parts of Brazil, Bolivia and Paraguay, to the mangrove swamp forests of Indonesia's Sundarbans, where tigers and monkeys live alongside fauna more conventionally identified with swamps, like crocodiles and crabs. They range from the swamps of Africa's tremendous, largely pristine Okavango Delta to small, dank pockets between suburbs and strip malls in New

Vasyugan swamp,
Siberia.

Jersey or Florida, stubborn survivals from a near-forgotten age. Even defined strictly according to prevailing, accepted scientific terms, swamps are, if not ubiquitous, scattered the world over, and richly varied in flora and fauna.

Whether we regard swamps as a specific subset of wetlands or as a general catch-all term for damp, sodden regions will vary depending on historical and cultural context. The important idea for this book, though, is that until very recently 'swamp' was an inherently inexact term, applied to spaces that themselves defied easy categorization as land or water, solid or liquid. The *Oxford English Dictionary* defines swamp as 'A tract of low-lying ground in which water collects; a piece of wet spongy ground; a marsh or bog. Orig. and in early use only in the N. American colonies, where it denoted a tract of rich soil having a growth of trees and other vegetation, but too moist for cultivation.' In the North American colonies, English settlers confronted wild spaces that had been largely cleared in their home country, and had no term to describe them; they adopted 'swamp' and 'dismal' to describe them.[4] In common parlance, swamp, marsh and bog are often interchangeable.

The indeterminacy of swamp-related language comes through clearly in terms widely regarded as interchangeable with 'swamp'.

Beautiful blue-eyed
royal Bengal white
tiger swimming in
an algae lake.

Functioning more as a reflection, of course, of how we use language than as a prescriptor, *Roget's Thesaurus* lists the following synonyms for swamp: 'bog, bottoms, everglade, fen, glade, holm, marsh, marshland, mire, moor, morass, mud, muskeg, peat bog, polder, quag, quagmire, slough, swale, swampland'. The terms take us on a virtual tour of widely divergent locales: 'fen' might bring to mind New (or Old) England, while 'muskeg' resonates with colder, more northerly climes like those of Alaska and northern Canada. 'Moor' might call up images of the English countryside, while peat bog evokes, perhaps, Ireland – all well removed from the sultry, moss-canopied swamps of Louisiana or Virginia. At times, this book will treat the swamp with some precision, adhering to its specific and proper definition; when appropriate, it will range beyond the proper swamps to wetlands writ large, visiting fen and marsh, moor and bog.

The capybara, common in the Brazilian Pantanal.

Before we leave Roget behind, consider the more connotation-loaded terms in this list: 'mire', 'morass', 'quagmire', 'slough',

A moor in Germany's Lüneburg Heath.

unpleasant terms offset by the presence of the more picturesque 'everglade' and 'glade'. Through the felicitous happenstance of alphabetical order, there is a kind of accidental poetry to this list, as it carries us through a tour of types, tropes and attitudes. We slog through bogs and bottoms, pause in glades and muck through morasses and quagmires before we culminate in the all-inclusive finality of 'swampland'. Swamp has meant, and continues to mean, all of these things and more, and remains a space of fascinating contradiction.

Defining wetlands is more than a semantic technicality. As wetlands' ecological importance and fragility have been gradually recognized, efforts at preservation have made the legal question of what exactly constitutes a wetland a matter of acute practical importance. Legal definitions can be freighted and highly contentious, as property values and potential for development can hinge on whether or not a given space is technically

a wetland. Establishing whether a particular space is legally a wetland, for example, may determine whether it will be preserved or drained, filled in and paved over, at least in countries with legal provisions for preserving wetlands. A century and a half ago in the United States, legally defining swamplands meant determining which lands would be targeted for draining to improve productivity and public health, and became a point of great contention. For example, as United States lawmakers grappled with the question of how to implement the Swamp Land Acts in the mid-nineteenth century – laws that turned over federally owned swamplands to the states, with the aim of encouraging their drainage and development – they found that their definitions of 'swamp' varied widely from state to state. One frustrated senator from New Hampshire complained of the shifting definition of swamplands in the bill's language: 'There is nothing to ascertain what these lands are. There is no criterion given; the whole is loose and indefinite, and uncertain!'[5]

The interplay of the two terms 'swamp' and 'wetland' captures the ambivalence and changing attitudes towards these areas rather vividly. To many throughout history, swamps have been dark, dangerous, even infectious places, full of strange, dangerous beasts, both natural and supernatural. All but the savage or desperate shunned their boundaries. Until recently, they have generally been regarded as problems to be solved, areas to be cleared and drained in the name of civilization and progress. Wetlands, on the other hand, are essential to life. They clean and filter water, nurture biodiversity, provide natural flood protection and maintain coastlines. They have been imagined and depicted as havens of purity and authenticity, untouched by the pollution of progress, and those who dwell in them have often been depicted as noble aborigines or as guardians of authentic, fading cultures. These contradictory representations often blend into one another in ways befitting the blended ambiguity of the lands they describe.

Primordial, mysterious, antithetical to civilization, swamps are daunting spaces. Since swamps are among the wildest of wilderness spaces, and among the most resistant to mastery and

A bog island in
Europe.

cultivation, peoples' and cultures' attitudes towards them reflect a
great deal about their attitudes towards nature as a whole. These
attitudes are often fundamentally conflicting, if not paradoxical.

Through much of history, swamps were considered foul,
poisonous places. Swamps were believed, essentially, to breathe
corruption into the air, emanating miasma that carried infectious
disease born of rot and putrescence. For much of history, malaria
was believed – as its name, Italian for 'bad air', indicates – to
originate in the vapours of swamps themselves, until a scientist
named Alphonse Laveran traced its actual cause to the parasite
Plasmodium, transmitted by mosquitoes that thrive in swampy
areas.[6] The 'breath' of the swamp, the miasma whose name comes
from the Greek term meaning to pollute or to stain, was long
regarded as a cause of sickness, death and general blight. Underlying the various superstitions, legends and lore about devils,
witches and supernatural terrors in the swamp was, it long
seemed, a legitimate terror that could poison with a single breath.

Ironically, however, swamps actually act as purifiers rather
than polluters. Swamps filter pollutants from water. In fact, the
waters of the Great Dismal Swamp were legendarily pure: the
tannins that leeched into the water from the bark of its trees

kept bacteria from growing even as they dyed the water an amber colour, and sailors before the coming of refrigeration carried barrels of swamp water and praised its healthy, even mystical qualities.[7]

Alligators sharing a log.

Swamps have always been frightening – threatening. They have been practical problems, obstacles to settlement and progress, threats to order, and challenges to ideologies that emphasize human dominion over nature, both practically without and symbolically within. They have been havens for beasts – poisonous snakes, snapping turtles, alligators – as well as for refugees, criminals, escaped slaves and society's rejects. Folk tales and legends populate them with witches, spirits and monsters – at times, they are even called 'Devil's domain'. Dark, foreboding, sublimely primal, swamps have long been synonymous with mysterious dangers and occult evils.

While these associations have never quite gone away, and still imbue contemporary swamps with mystery and fascination, since the nineteenth century swamps have come to be perceived, for ever more compelling reasons, as fragile, threatened and endangered. In a zero-sum game of nature and civilization, the pastoral swamps are always shrinking in the face of progress. One thing that is certain about wetlands is that they were once much more widespread and plentiful than they are today. The last century, in particular, saw rapid destruction of wetlands, through climate change and human development and drainage. Sadly, very real threats to the world's swamps are ubiquitous and growing. Louisiana's coastal wetlands are vanishing at an alarming rate, and even such sparsely populated areas as the Okavango Delta and Brazil's Pantanal are threatened by urbanization and development at their edges. Bogs, too, are imperilled: over 75 per cent of England's peat bogs have been destroyed by development and peat extraction. Swamps tend to house endangered animals: the Vasyugan Mire is home to many endangered bird species, and in the Sundarbans of Bangladesh and India, tigers face severe threats to survival. In most places, the swamps that once seemed so menacing and indomitable are now preserved through law and custom, threatened by pollution, natural disaster, development and climate change.

Foggy winter morning near Bayou Benoit, Louisiana, December 1992.

These oppositional ways of thinking about swamps are sometimes morally charged. Swamp-dwellers have often been criticized as slothful, indolent, content to live off the land rather than to master and cultivate it. In more extreme cases, swamp-dwellers, assumed to embody the wildness of their home, have been labelled savages. How that savagery was imagined depended, yet again, on perspective; what Puritans saw as demonic wildness, poets might celebrate as prelapsarian purity.

Swamp lore and legend are rife with images of the under-world, of ghosts and spectres, of death and undeath. Ironically, though, wetlands usually teem with life, and in fact are often essential to the survival even of species that generally live out-side their borders. Part of the slothful stigma often attached to swamp-dwelling came from the ease with which at least a

Birds eating at the Pantanal, Brazil.

Jaguar on a riverbank in the Pantanal, Brazil.

subsistence could be had from the swamps' bounty with relatively little effort, which represented an affront to the entrenched work ethics of observers.

The swamp has always been rife with paradox, and is perhaps particularly suited to it. With the introduction of their opposites, prior conceptions of the swamp do not fade; rather, they coexist, paradoxically, alongside their antitheses – mingling, even as land and water mingle, even as the pragmatic and the fantastic mingle, in the physical and cultural space of the swamp.

These uncertain, ambiguous, liminal spaces have been targeted for drainage and designated for preservation, imagined as evil, savage and pestilential and as pure, pristine and unsullied by civilization. They filter water, control flooding and teem with wildlife, infinitely varied and sometimes dangerous to humans. Swamps have always been subversive spaces, resistant to dominion, draining, cultivation; cultural attitudes towards swamp spaces provide a clear index of attitudes towards the natural world writ large. This book considers swamps not only as physical realities, but as ideas. In other words, understanding swamps entails grasping not only what they are, but what they mean – what they have meant at different times and to different people. This is what makes them fascinating.

1 Swamp as Home: People of the Swamps

Before the palaces of Ur were built men had stepped out from such
a house and launched canoes like this to go hunting in the reeds . . .
Five thousand years of history were here, their pattern of life little
changed.
Wilfred Thesiger

The Garden of Eden, some say, lay within a swamp.

Historians' efforts to locate a possible site for the biblical
birthplace of humanity often lead them to southern Sumer,
among the marshes of southern Iraq.[1] Further, Sumerian and
Babylonian Creation legends evoke images of a world conjured
from swampy chaos. The Enuma Elish, an epic poem probably
handed down from even more ancient Sumerian tradition before
being transcribed in 2000 BC, describes the world's origins as the
outcome of a desperate struggle between the gods of Good and
Evil, of Order and Chaos. Once Marduk, also called Enlil, the
chief god of the Babylonian pantheon, routed the dragons and
serpents of the army of chaos, he made the sky, the stars and
eventually the world. Gavin Young, a journalist and travel writer
who lived among the Marsh Arabs of Iraq for several years in
the mid-twentieth century, draws a direct connection between
Marduk's creation of the world and the Marsh Arabs' creation
of their wetland homes: "'He built a reed platform on the surface of the waters, then created dust and poured it around the
platform'– and this, briefly stated, is how today's Madan [Marsh
Arabs] create the artificial islands on which they site their reed
houses.'[2] Whether or not we embrace biblical or other mythological origin stories, it is clear that human civilization began in,
and eventually emerged from, wetland spaces.

Traditionally, swamps have been associated more with the
absence of humanity than with its origins. Indeed, we often
define swamp spaces as wilderness, either implicitly or explicitly

outside human dominion. Given the stigma attached to swamps and bogs, it may seem surprising that so many of the world's civilizations began and developed in or near wetlands. From Mesopotamia to the Mekong Delta, from the Macedonian marshes in the era of Alexander the Great to the Niger Delta in Mali, from Rome by the Pontine Marshes to the Netherlands, wetlands have seen the birth and development of civilizations all over the world. The Fertile Crescent, the area between the Tigris and Euphrates rivers, is characterized by abundant fresh-water swamps that support tremendous biodiversity. The area is generally acknowledged as the cradle of human civilization: our earliest evidence of both written language and of the development of foundational technology such as the wheel comes from this region. Archaeologists surveying 'wet sites' all over the world have uncovered evidence of settlements both among and on the edges of swamps, marshes, fens and bogs, dating back to prehistory.

Early peoples probably settled among the wetlands for a variety of reasons. Wetland agriculture offered advantages in that it required little in the way of technology, offering relatively comfortable subsistence. Rich in wildlife and fertile, swamps and marshes provided sustenance as well as safety to early peoples. Wetland waterways allowed people to travel easily by boat before land travel by road was a widespread possibility. Peat bogs offered little in terms of arable land or easy travel, but provided fuel for burning and material for building, two essential elements for human settlement. The natural bounty of the wetlands made them attractive, if not for people to settle among them, then at least for them to build settlements along their edges.[3]

Archaeologists have discovered evidence of prehistoric wetland settlements all over the world. In Japan, the Kamo settlement gives evidence of a group of hunter-gatherers who lived at the edge of the swamp, hunting deer, otter and wild boar, and fishing for dolphin and other sea life during the Jōmon period, an era of Japanese prehistory that began around 14,000 BC. In western England, the Somerset Settlement sites at Glastonbury and Meare provide evidence of late prehistoric people who farmed the wetlands to develop prosperous, sheltered homes in the midst of the marsh. Similar settlements throughout Europe, Africa, Asia, the Americas and all over the world tell similar stories of early peoples who prospered in wetland or wetland-adjacent settlements, cultivating rice and other staples suited to wetland agriculture, and supplementing them with swamp fowl, game and fish.

For those who did venture into their depths to settle, swamps and marshes also offered protection and concealment. Archaeologists have discovered evidence of such protected settlements all over the world. In New Zealand, archaeologists have discovered versions of Maori *pa*, fortresses often associated with hilltop settings, adapted to swamps and crannogs. These fortresses, built by filling a framework of partially buried stakes with gravel and rock to form a foundation and enclosing the structure with a fence, could typically only be reached by boat. They were fortified against attack, surrounded by sharp submerged stakes that could capsize the unwary and were evidently used throughout the prehistoric era in Aotearoa.[4]

We are able to learn a surprising amount about people who lived so long ago because of the remarkable troves of archaeological knowledge found in wetland sites. These sites are extraordinarily important from an archaeological standpoint because materials like wood, cloth and even – or especially – human bodies can be preserved in peat and mud in ways that are not possible in other environments. Many ancient peoples around the world, even if they did not live among the wetlands proper, used swamp, bog and marsh areas as burial grounds. Ancient cemeteries like the one at Windover in Florida indicate that early Native Americans used wetlands this way dating back to around 5000–6000 BC. At Windover, bodies have been found that were wrapped in mats made of grass, then submerged in ponds and held to the pond's bottom with stakes. As the pond filled with peat, the bodies were preserved. Sites like this one, as well as the bogs of northwestern Europe, from which a great many bodies have been unearthed, indicate that ancient people

Amazing Pantanal River – the Pantanal in Brazil is one of the world's largest tropical wetland areas.

in many places around the world used wetlands as burial places, likely because of their potential preservative properties.[5]

Bog bodies are a unique phenomenon that offer incredible glimpses into the very distant past. The earliest discovery of a bog body dates back to 1773, while the first photograph of one was taken in 1871. A distinctive combination of factors, including temperature, acid produced by the sphagnum in the soil and protection from air and from organisms that might feed on corpses, gives bog peat astonishing preservative properties. Researchers have unearthed bodies from Windover that have been preserved almost perfectly – flesh, clothing, brain matter, teeth, even stomach contents – for more than 7,000 years.[6] Bodies have also been unearthed from many sites in Europe. The famous Grauballe Man, a magnificently preserved body found in a peat bog at the village of Grauballe in Jutland, Denmark, dates back to the third century BC. He and Tollund Man, another

Wet peat on a bog, Torronsuo National Park, Finland.

bog body found in Jutland and dating back to the fourth century BC, were both unearthed in the 1950s, with soft tissue, clothing and internal organs intact. Another peat bog near Lindow Moss in the northwestern part of England yielded two famous bog bodies – first Lindow Woman in 1983, then Lindow Man, humorously nicknamed 'Pete Marsh', in 1984. Bodies like these provide a wealth of information about the past, but also lead to mysteries. Many bog bodies show evidence of tremendous violence. Their throats have often been cut, cords or ropes are found tied around their necks, and their bodies show other signs of intentional, perhaps ritualized violence. The dark stories hinted at by these violent deaths tend to enhance the aura of sinister mystery surrounding bogs.

Wetland archaeological finds also indicate that wetlands were sacred spaces to ancient peoples. Votive effigies dating back to the Gallo-Roman era have been unearthed from the bottoms of spring-fed pools in France. Sites in New Zealand, Finland and Ireland, among others, have yielded prehistoric treasures that strongly suggest sacrifices of worldly goods to appease or to appeal to entities associated with wetlands.

However welcoming and sacred wetland spaces may have been for prehistoric peoples, a general pattern seems to hold true nearly everywhere. As people developed technology and were able to exert more control over their natural environments, they tended either to leave marshes and swamps behind, or to drain or clear them. Indeed, one vision of European history, as summed up by the anthropologist Stuart McLean, frames it as 'one of protracted struggle against the abject, waterlogged expanses marking its own inner and outer frontiers'.[7]

Indeed, in most places around the world, the story of human interaction with wetlands is one of gradual modification and eventual control through the development of specialized agriculture and technology. The building of a superstructure of roads over treacherous bogs to connect one island to another; the introduction of irrigation and drainage systems in marshes and swamps – signs of ancient peoples modifying and repurposing wetland spaces to fit their needs have been found all over the world. In Bronze Age Britain, people appear to have inhabited the coastal marshes, taking advantage of marsh agriculture and the salt they could extract from the briny waters, but by the Iron Age, most had moved inland. In the Netherlands, coastal wetland dwelling persisted considerably longer, as inhabitants mastered techniques of wetland agriculture. In fact, the effectiveness of their processes of taming the wetlands led to the comparatively rapid evanescence of the wetlands hundreds of years later. A general pattern emerges when studying inhabitants of coastal wetlands in the period stretching from the Early Neolithic era to the end of the Iron Age. Swamp- and marsh-dwellers tended to move from living off the bounty of the land itself to cultivating and modifying it to better suit their purposes. As their techniques became more specialized and sophisticated, they typically moved further inland; as they became more adept at taming the land, they became less dependent on the natural environment. Indeed, in many places and for many cultures, the story of civilization is a movement out of the swamps and marshes, followed by systematic draining or taming of or otherwise dominating them.

In the ancient world, many cities and towns were built in close proximity to marshes and swamps. This nearness brought with it both advantages and disadvantages, both in terms of practical concerns and philosophical notions about the nature of civilization. The ancient Greeks and Romans prized order, and saw aesthetic appeal in the tamed, controlled landscape. Swamps and marshes, then, were fundamentally offensive to this ideal. Ancient texts by classical writers like Vitruvius and Columella warn against the hazards of settling near miasmatic bogs and marshlands, for reasons ranging from the abundance of snakes and insects to the corruption of their dampness itself.[8] Nevertheless, cities like Ravenna, Alexandria, Agrigento, Babylon, Syracuse and even Rome itself rose up among swamps and marshes, and enjoyed a measure of tactical advantage due to their defensive properties.[9] People have regarded swamps and marshes with profound ambivalence throughout most of human history.

Modern footpath or gangplank over a pond in the woods. Old trees standing in a moor or swamp in the forest, with sunbeams and smooth light falling through the treetops.

Despite their ubiquity and frequent proximity to urban areas, the swamps' primal associations come in large part from their resistance to cultivation and development; most often, 'swamp peoples' are characterized as, if not aboriginal, markedly less 'developed' than others. Whether Europeans coming from the bogs to build civilizations around hall-centred settlements, or African tribes draining and cultivating swampland to advance and improve agriculture, a culture's emergence from the swamps is generally regarded as a step in its development and modernization. In some senses, such a designation may be colonialist denigration or cultural chauvinism; often, though, it reflects a harsh reality of limited access to medicine, education and contemporary technology. Because most swamp-identified cultures have not represented themselves through written records, the understanding we have of them is inevitably filtered through the perspectives of outsiders. These outsiders may be visitors or newcomers, viewing the swamp denizens from the perspective of a colonist or tourist, or they may be descendants, trying to connect with a vanished or vanishing cultural tradition, blending genuine cultural tradition with nostalgia. Just as the swamps themselves have been regarded, constructed and interpreted in myriad ways, so have those who dwell within them. Romanticized and demonized, scorned and elevated, depicted as paragons of virtuous simplicity and as dangerous savages, swamp people have existed at the periphery of dominant culture, and have often become screens on which that culture projects its values, personifies its fears and creates its history.

Unquestionably, certain consistent stigmas have been attached to peoples who have remained among the swamps after others have moved beyond them. Swamp, marsh and bog people have a long history of being denigrated by judgemental outsiders for their backwardness, laziness, lack of industry and general lack of development. The Roman naturalist and historian Pliny, writing in the first century AD, perceived the inhabitants of coastal wetlands in the Netherlands and Germany as 'a wretched race', living among the inundated marshes like 'so many shipwrecked men'.[10] Pliny lamented that even these miserable beings would not appreciate the deliverance and redemption of their conquest by Rome – an obvious rescue from savagery, to his eyes. In the nineteenth century, the English historian John Lothrop Motley, composing his *History of the Rise of the Dutch Republic*, explicitly frames his work as recounting 'the gradual development of what is now the kingdom of Holland, from a race of ichthyophagi who dwelt upon mounds which they raised like beavers above the almost fluid soil'.[11] The reactions of outsiders to swamp and wetland peoples inevitably mirror their attitudes towards the primitive, and are often predictably polarized, tending towards the denigration of the 'savage' or elevation of the pure and pristine. And because few swamp-identified cultures write or speak for themselves – at least, not until after they have largely joined the modern world enough to assert themselves as part of the conversation – those who wish to learn about them must be attuned to enhanced levels of the same biases that colour all history written by outsiders, whether colonialist, Romantic or otherwise. This chapter will explore a variety of swamp-dwelling peoples of the past and present, examining as it does so the intermingled associations of landscape and culture. The stories of these various peoples hinge, in particular, on the level and nature of their contact with what we might call the Outside World – peoples outside their often intentionally insular and sheltered cultures.

While the vast majority of swamp-dwelling peoples have entered the modern world through some combination of coercion and choice, a few isolated peoples remain who live among the swamps in much the same way that prehistoric tribes would have done. Papua New Guinea, one of the few largely uncharted places left on earth, is populated by a wide array of small, disparate tribes, some of which have had very little contact with the outside world. Among the area's forested swamps, which make up around a third of the area's total land mass, tribes persist who have adapted creative ways of living in their swamp environments. The most remote of these tribes live as hunter-gatherers, and some have not even made the shift to metal tools, still using stone axes and bamboo utensils. They have, however, developed remarkable ways of adapting their swamp environments to suit their needs.

The area's swamp-dwelling tribes took to the swamps, most believe, as a haven from a headhunting tribe known as the Marind-Anam. These warriors were so fearsome that Dutch colonizers named two local rivers in their honour: the Moordenar, which translates as 'Murderer', and the Doodslager, or 'Slaughterer'.[12] Consistent with a pattern among many contemporary swamp-dwelling peoples, less warlike or powerful

groups fled into the swamps for protection and refuge. Nick Middleton, an English explorer, geologist and celebrity survivalist and author of *Surviving Extremes: Ice, Jungle, Sand and Swamp* (2003), describes the alienness of the remote swamp environment to his outsider's eyes:

Two men of the Kombai tribe sitting on the veranda of their treehouse territory, Indonesian New Guinea.

> Only in the twentieth century did outsiders seriously begin to explore New Guinea's interior, slowly lifting its cloak of mystery to glimpse a world still living a prehistoric existence. It was the stuff of storybooks, a virgin land populated by birds of paradise and people armed with bows and arrows whose cultures centered on a ritual cycle of head-hunting. To the outside world, New Guinea was the epitome of Beauty and the Beast.[13]

Middleton describes encounters with two swamp-dwelling tribes, each of whom has adapted to its environment in distinctive ways. The people of the Jeobi village, a settlement located in the heart of a swamp, create their own islands by cutting and floating grasses to form great mats that float on the water's surface. The Jeobi people, though they may have initially been driven into the swamps by the threat of headhunters, now live in the swamps by choice, having rejected the offers and requests of missionaries and government alike to move to higher ground. Indeed, in the 1970s the government built an entirely new village for the Jeobi upstream, out of the heart of the swamp, reasoning that with the threat of headhunting now in the past, the people could begin to embrace a more modern, or at least less strenuous and isolated, way of life. Almost all of the villagers returned to the swamp within a few years.[14]

Middleton's take on the Jeobi village is similar to that of many outside explorers coming upon swamp-dwelling peoples living in ways that have changed very little since prehistory. While he makes no bones about the awfulness of swarming mosquitoes or the danger of the swamps' crocodiles, he sees the picturesque floating islands of the Jeobi as 'a tropical Venice with coconut palms and dug-out canoes . . . It was the most unexpected of things: a little piece of paradise in a swamp.'[15]

The Kombai, another tribe dwelling in the swamps of Papua New Guinea, have developed a very different strategy for coping with the swamp environment. This hunter-gatherer tribe live in tree houses many metres above the ground. Safely ensconced among the treetops, the Kombai escape the treacherous swamp ground, the clouds of mosquitoes and much of the heat of the swamps. Cooled by breezes, the huts are easily defensible against invaders, as the climbing poles that allow access can be easily pulled up. The Kombai and Jeobi live in a manner that has remained largely unchanged for centuries, and provide glimpses of how ancient peoples may have interacted with the swamps before technology allowed for their drainage and taming.

The very fact that we know about these peoples, however, indicates that their way of life is endangered. Magazine and television stories have been run in recent years about the Kombai; Nick Middleton's work yielded an adventure travel series on the BBC that garnered international attention. The forests and swamps of Papua New Guinea, like those almost everywhere, are being consumed by industry, logging and development. These swamp-dwelling peoples, described by outside chroniclers in terms that emphasize the purity, altruism and unspoiled nature of their cultures, may finally be facing the cultural transformation from which the swamps have thus far largely shielded them.

Perhaps the archetypal swamp people, the Marsh Arabs of Iraq, also known as the Ma'dan (or Mi'dan), provide a model for how a relatively isolated, insular group might persist largely unchanged for centuries, if not millennia, only to have their way of life forever altered as the forces of modernity inevitably overtake them. The Ma'dan have lived among the Mesopotamian marshes for thousands of years. Many regard them as descendants of the peoples of ancient Sumer and Babylon, and believe that they may represent an unbroken connection to the world's most ancient civilizations.[16] The marshes of the Fertile Crescent represent the Middle East's most extensive wetland ecosystem. Flooded periodically as the Tigris and Euphrates rivers overran their banks, the marshes once spanned between 15,000 and 20,000 square kilometres, and were home to a wide array of bird, fish and mammal species, as well as host to millions of migratory birds each year until the late twentieth century, when war, exploitation and, eventually, concerted efforts by Saddam Hussein's regime to drain, clear and even poison the marshes largely destroyed them and the people who had lived among them for millennia. Now, both marshes and people have very nearly been eradicated.

Like many swamp- and marsh-dwelling peoples, the Ma'dan have traditionally been a fairly insular people, cut off from the outside world by a combination of choice and circumstance. Most of the glimpses into Ma'dan culture that have been afforded to the outside world have inevitably been filtered through the eyes of outsiders – travellers, explorers, anthropologists. Perhaps the most notable visitor and chronicler of Ma'dan culture was the famous British world traveller, explorer and travel writer Wilfred Thesiger, who spent seven years in the marshes of Iraq in the 1950s. Thesiger had a pronounced anti-modern streak: he resented the adulteration of pristine and natural areas with automobiles and other trappings of civilization, and praised the marshes for having 'the stillness of a world that never heard an engine'.[17] Thesiger saw in the Ma'dan direct descendants of ancient peoples, virtually unchanged in 5,000 years, hidden away from what he called 'drab modernity with its uniform of second-hand European clothes, which was spreading like a blight over so much of the world'.[18]

Gavin Young, a companion of Thesiger's who also lived among the Ma'dan in the 1950s, waxed poetic about the timelessness of the marshes, which seemed to him, at the time, infinite and eternal:

> An aura of infinity hangs over these sometimes exhil-
> aratingly beautiful Marshes, these sometimes gloomy
> and disturbing 6,000 square miles of water and reed.
> And why should it surprise, that intimation of infinity?
> Is it a little matter that five thousand years ago the kings
> of Ur of the Chaldees gazed at the curved reed houses
> that we can gaze at and visit today? That we can travel
> today in the royal gondolas of Sumer and Babylonia?[19]

Thesiger and Young saw in the Ma'dan a connection to time immemorial, dwelling in marshes that in the 1950s were largely as they had been for millennia. The Ma'dan that they encountered were Shiite Muslims, descended from Arab tribes dating back thousands of years. Driven into the marshes by invading Mongols, the Ma'dan settled among them, fishing, keeping buffalo, growing crops like rice and crafting homes, boats and even islands from the reeds and rushes surrounding them. The Ma'dan's way of life developed thoroughly in response to their environment. They constructed their floating houses and small man-made islands from marsh reeds. As Young describes it,

You make one [house] just as Marduk made the world.
You decide how big you want your house to be. You gather
a small mountain of rushes and heap them in the water
inside a reed fence that rises above the surface. When the
well-trampled mass of green also appears above the surface,
you fold the fencing in on top of it; and continue piling
and stamping reeds until you are satisfied with the size
and compactness of the new island you have just created.[20]

The centuries-spanning history of the marshes and their inhabitants is a story of irrigation systems built and destroyed, of the ebb
and flow of civilizations, and of the coming of Arab refugees from
various conflicts and wars. As the buffalo-herding, agriculture-
focused Ma'dan mingled with the incoming tribesmen, who were
often more warlike, the Ma'dan became fighters as well as farmers. Though the marshes had been home to rebels and outlaws
since the very beginning of their human occupation, the Ma'dan's
traditions, habits and patterns of behaviour all came, Thesiger
asserts, from the 'code of the desert Arabs'.[21] They were simply,
and happily, cut off from the rest of Iraqi culture by the marshes
in which they lived.

The marshes had offered sanctuary and protection since
the earliest times. The Chaldeans used the shelter of the thick
reeds, which could only be traversed by small watercraft, to their
advantage as they defeated Sargon, ruler of the Assyrians, in
the seventh century BC. In 705 BC, King Sennacherib declared
himself ruler of all the world, and set out to subjugate Babylon.
He and his men took Babylon, but the king, Merodachbaladan,
escaped into the marshes. Taken in and helped by the Ma'dan, he
hid successfully, thwarting Sennacherib's efforts to find him: as
Sennacherib recorded, 'I hurried after him and sent my warriors
into the midst of the swamps and marshes and they searched for
him for five days, but his hiding place was not found.'[22]

Indeed, through most of the Ma'dan's history, they lay
beyond the control of any sort of central government or authority. The marshes still harboured criminals and refugees, but
also an independent and self-sufficient people who could not

be subjugated. Even the mighty Ottoman Turks were unable to assert their rule over the Ma'dan, who remained under the authority of local sheikhs during their reign, and 'completely ignored the Ottoman governor of Baghdad'.[23] By the time of Thesiger's writing, the Ma'dan appeared equally untouched by central authority: 'As for Iraqi officials, I felt certain that none of them had been further into the Marshes than was absolutely necessary.'[24] Here, then, were a self-sufficient, self-sustaining, fiercely independent people, living in the way their ancestors had lived for centuries.

It may be tempting to romanticize the life of the Ma'dan, as Thesiger and Young sometimes do, by envisioning a natural, sustainable world shielded from the polluting forces of modernity. Even before the late twentieth-century campaigns that effectively destroyed the marshes and routed those who lived among them, the Ma'dan faced an array of challenges uniquely linked to their marsh environment. Clouds of mosquitoes and other biting insects were endemic in the marshes, as were disease-bearing parasites. Edward Ochsenschlager, an ethnoarchaeologist who lived among the Mi'dan (as he calls them) for years between the 1960s and 1990s, reports that bilharzias, a disease carried by flatworms that can enter through any break in the skin, was nearly universal in his area:

> few people in the area were free of this disease, with which, although curable, they were often re-infected each time they stepped into the marsh or canal . . . Like people everywhere, local inhabitants took their problems for granted and found them of minor consequence. Few attributed blindness, external bleeding, and serious internal problems to the flatworms that caused bilharzias.[25]

Another bane to the Ma'dan were the enormous wild boars that populated the marshes. Larger than boars in most places, these wild pigs devoured crops and often killed people who encountered them. The comparative downplaying of the marshes' hazards in famous accounts like Thesiger's and Young's largely reflects the perspective of the intrepid adventurer. Thesiger embraces the harsher elements of marsh life: 'Each night as I lay down to sleep a cloud of mosquitoes settled on my face, and a weight of fleas moved under my blankets, but I accepted this as a small price for the contentment I had found.'[26] Thesiger's perspective on more dire threats, like that posed by the wild boar, is explicitly at odds with that of the Ma'dan themselves: 'The Marshmen hated the wild boar, their only natural enemy after lion were exterminated following the influx of modern rifles in the First World War. I remember an old man saying, "Pigs! They are the foe. They eat our crops and kill our men. God destroy them!"' Thesiger, though, admits that he relished the presence of these dangerous creatures: 'Their massive dark shapes, feeding on the edge of the reed-beds at evening, were for me an integral part of the marsh scene. Without the constant risk of encountering them life for me would have lost much of its excitement.'[27]

The larger threats to the Ma'dan, though, came from without, not within, the marshes. Even before the focused campaign that drained the Iraqi Marshes in the 1980s, their culture was under siege by a variety of forces, including the large-scale conscription of young men into the military. Many did not return from the Iran–Iraq war, leaving families to suffer or starve; those who did return brought with them outside ideas and ideals, learned in their interactions with fellow soldiers from other areas. Radio, the importation of goods and a gradual shift from a barter system to a cash-driven economy had all begun to invade the insular world of the marshes in the 1960s and '70s. Ochsenschlager also reports that, by the 1980s, relations between the Ma'dan and the Beni Hasan, who had lived on the marsh edges and coexisted peacefully with them for ages, had begun to deteriorate. The Beni Hasan began to speak of the Ma'dan in ways long associated with marsh and swamp-dwellers: 'Occasional allusions, in passing, characterized the Mi'dan individually or collectively as dirty, lazy, venal, and not too bright . . . visiting officials would regularly denigrate the Mi'dan and the way they made a living.' Even as Hussein's campaign to drain the marshes proceeded, elsewhere, the Ma'dan in Ochsenschlager's area were driven out by environmental and cultural changes as well as 'a barrage of propaganda alluding to their unprincipled corruption'.[28]

A concerted and multinational effort to restore the Iraqi Marshes is underway, and is making some progress. The future for the Ma'dan and their culture, though, looks bleak. The contemporary Ma'dan who attempt to return to what remains of the marshes have none of the basic amenities necessary for contemporary life: they lack clean, potable water, power, public works, hospitals, schools. The waters of the marshes are now largely polluted, and residents who drink them have high rates of dysentery and other diseases. Moreover, the insular culture that enchanted Thesiger and Young has been forever altered, if not destroyed, by a combination of concerted effort and the general siege of encroaching modernity. The Ma'dan are, in this sense, an archetypal swamp culture.

Other swamp peoples experienced this clash in different places and historical eras. In the North American colonies, a clash of cultures shaped the destiny of swamp-dwelling Native American tribes. Terrified of wilderness as antithetical to God's order, European colonists in North America often identified 'savage' Native Americans with swamps. Many of these observers, their perspectives influenced by Puritan rhetoric dating back to the 'City upon a Hill', connected the 'godless savage' with howling wilderness, and assumed that for people defined by savagery, the wildest landscapes would be the most natural home. Native American attitudes towards nature, of course, differed starkly from this Puritan anxiety and distrust. Indeed, Native Americans of the pre- and early colonial eras most likely regarded themselves as fundamentally part of nature, rather than destined or commanded to conquer it.[29] Despite this fundamental difference, the truth of most Native Americans' attitudes towards swamps was more nuanced, and swamps were not entirely free of trope or stigma, at least among the Virginia Indians. A minister's report from Virginia, written in 1689, describes a vision of heaven and hell that tracks intriguingly with the Virginia landscape. While the good go to a place far away 'where there is no extreme heat, nor cold, nor storms, but the air alwayes clear and serene', the wicked

> wander up and down about their marishes . . . They say that this horrible wilde place, to which the wicked are condemned is in sight of the other, but they can never come at it, for the bushes and briars, and swamps and marishes, that are between them and it. They say that the wicked after this life are alwayes hungry and thirsty, but have nothing to eat and drink, but what is raw, as frogs, flies, etc.[30]

The age-old trope of the hellish swamp was not limited to European tradition.

Many Native Americans saw opportunity, rather than suffering, in the Southern wetlands. Further south, in Florida, tribes like the Calusa found freedom from need in the Everglades, a vast region of tropical wetlands in the southern part of Florida. Such groups drew their sustenance from the bounty of the wetland region without needing to devote much time or energy to active cultivation and agriculture. Evidence suggests that they created remarkable works of art and engineering projects before being wiped out, largely owing to imported European epidemics like smallpox and the bubonic plague, by 1900.[31] The ravages of such imported diseases were by no means limited to swamp-dwelling Native Americans; still, they invert the idea of the swamps themselves as poisonous and pestilential.

The tribe most indelibly associated with the Florida swamps in the popular imagination is the Seminole, who supplanted the Calusa in the Florida swamplands. The Seminole, a tribe formed of indigenous Florida Native Americans and others, most of the Maskókî tribes, were forced south from Georgia and Alabama over the course of the Indian wars of the late eighteenth and early nineteenth centuries. The Seminole's very name characterizes them in opposition to European invasion: an adaptation of the Spanish term *cimarrones*, or 'free people', it defined a group who resisted domination and removal. Alternately

'Indians Stalking Spaniards Riding Horses in the Swamp', frontispiece to John S. C. Abbott, *Ferdinand De Soto, The Discoverer of the Mississippi* (1873).

demonized and romanticized, the Seminole became linked with the swamp, for better or worse, in the popular imagination. A study of the various imaginative tropes by which whites characterized the Seminole provides a veritable catalogue of attitudes towards the swamps themselves.

Initial attitudes towards the Seminole mirrored early, practical attitudes towards swamps. As a wild, dangerous, warrior culture living in the untamed Florida swamps, they became, in the white imagination, inextricably identified with the landscape. In the late 1850s, after the end of the last Seminole war, the Seminole became a kind of bogeymen, their numbers uncertain – some claimed they were all but gone, while others claimed that as many as 1,500 remained, lurking in the Florida Everglades.[32] As the closing of the frontier robbed America of part of its essential mythology, the remaining pockets of wildness in the Florida swamps became new frontiers in popular culture, enabling pulp novelists to cast the archetypal frontier adventure of the brave white explorer venturing into the unknown and encountering the menacing savage in the Florida swamps rather than on the western frontier. The unknown, unquantifiable and indomitable swamps, populated by unknown, unquantifiable and indomitable Indians, provided a compelling surrogate frontier. Wild West literature like the popular magazine *Buffalo Bill Stories* transplanted western perils to the swamps of Florida, presenting Seminole as wicked savages who fed prisoners to alligators in sacrifice.[33]

Alongside such menacing representations was a powerful alternative depiction – no less an imaginative creation, but nearly antithetical to the demonic savages who populated many pulp westerns. Romanticism, with its redemptive, celebratory vision of the natural world, imbued the Native American with new symbolic significance. While the dominant depictions of Native Americans tended to fall into the old Puritan patterns of fear and demonization, the advent of Romanticism introduced an alternative vision that replaced depravity with nobility, the Devil with prelapsarian Adam. Writers like James Fenimore Cooper defined an American heroic masculinity that mingled conventional European values with Native American custom, dress and, to an extent, ecological consciousness. William Bartram, the famous Quaker naturalist who explored the southern landscape in the late eighteenth century, presented images of the Florida Seminoles that would help define the American Indian in the literary and popular imagination.

Influenced by his countercultural Quaker beliefs, Bartram brought to his depictions of Native Americans attitudes far removed from those that painted the Indian as a wicked savage: after an encounter with a Seminole in the swamp, he poses the question, 'can it be denied, but that the moral principle, which directs the savages to virtuous and praiseworthy actions, is natural or innate?'[34] Bartram's praise for the natural ethics of the Seminole would be echoed by nineteenth-century anthropologists, who focused on presenting them as alternatives to a rapidly urbanizing America fraught with unrest. Celebrating nature throughout his work, Bartram breaks from the tradition of stigmatizing swamp-dwellers as lazy to celebrate the ease with which the Seminole lives in the swamps:

> How happily situated is this retired spot of earth! What an
> elisium it is ... where the wandering Siminole, the naked
> red warrior, roams at large ... Here he reclines, and reposes
> under the odoriferous shades of zantholixon, his verdant
> couch guarded by the deity; Liberty, and the muses, inspiring
> him with wisdom and valour, whilst the balmy zephyrs fan
> him to sleep.[35]

To Bartram, the natives did not need European civilization, or
even religion – in fact, at one point, he observes that the Creek
hunt to excess, and blames it on the fact that 'the white people
have dazzled their senses with foreign superfluities.'[36]

Bartram's philosophical predispositions doubtless led him
to romanticize the conditions of the swamp-dwelling Seminole,
who had, after all, been driven into the swamps largely against
their will, and who, after Indian Removal began in 1815, existed
in pockets of resistance, using the forbidding swamps as protection against interlopers. Also, for all the swamp's bounty, it is
inaccurate to characterize the Seminole as living in luxury. Still,
as Clay MacCauley observes in an 1884 Bureau of Ethnology
report to the Smithsonian Institution, the fecundity of the South
Florida swamp soil and the mildness of the winters was enough
to free them from anxiety about the future, at least as far as crops
were concerned:

> I am under the impression ... that they do not attempt to
> grow enough to provide much against the future. But, as
> they have no season in the year wholly unproductive and for
> which they must make special provision, their improvidence
> is not followed by serious consequences.[37]

For those keen to find a pure, prelapsarian alternative to a rapidly
modernizing world, such freedom from worry was easy enough
to exaggerate into self-sufficiency in harmony with nature.

The decades following the Civil War saw other images of
the Seminole emerge. Rather than remaining contrasts to white
American civilization, the Seminole began to be co-opted as

Asseola, a Seminole
leader, *c*. 1842,
lithograph.

symbols for white American causes. Some embraced a mythologized version of the Seminole leader Osceola, whom legend
had provided with a black wife who had been unjustly enslaved,
and thus transformed into 'a symbol of devotion to freedom and
detestation of slavery'.[38] Others, worried by the influx of immi-
grants pouring into the country, adopted the Seminole, ironically,
as a touchstone of national identity: 'By using the Seminoles as
symbols of American bravery, whites constructed a national distinctiveness designed to protect American identity from foreign
influence.'[39] Eventually, of course, the image of the Seminole,
the embodiment of heroic resistance to conquest, became so
thoroughly co-opted that, to most Americans, Seminole calls to

Everglades Wilderness
Preserve.

mind a cartoon figure on a Florida State football helmet. As early as the turn of the twentieth century, a strong note of nostalgia entered into depictions of the Seminole, as

Seminole Indian and family in dugout canoe, Miami, Florida, c. 1910–20.

> ideas of the 'dying Indian' took on a wistful quality, especially for those Americans who believed that societal change had come too rapidly . . . Imagined as a 'pristine' example of America's indigenous past, the projected demise of these Natives became a heartrending end to a chapter in American history.[40]

This evolution, such as it is, roughly mirrors that of the swamp: as practical dangers fade, it is reimagined, rehabilitated, cele-brated, commodified and, ultimately, eulogized. Consider Charles W. Smith's musing about the future of the Seminole in an article in the *New York Times Magazine* in 1925: 'for each child born five of the older ones pass away into the "Land of the Sky." Soon they will have vanished from the earth. But they will have gone with heads held high, conscious that they have never been conquered.'[41]

Of course, not all swamp-dwellers were indigenous peoples. From the colonial era well into the twentieth century, swamp-dwellers in the United States have borne a strong stigma. For reasons already explored, both superstitious and practical, the swamp landscape was typically regarded as such an abysmal morass that the values and virtues of those who would live there were automatically in question. Who – what white person – would live in such a place? Why? Speaking out against the prospect of allowing Florida to join the Union as a slave state, John Randolph of Virginia – a state with its own share of swamps – decried it as 'a land of swamps, of quagmires, of frogs and alligators and mosquitoes!' Randolph added that 'A man ... would not immigrate into Florida ... no, not from hell itself!'[42] Native Americans living in the swamps were one thing – categorically different enough from the white man that they could safely be troped as savage. White settlers living in the swamps were another. They bore the same accusations of sloth and indolence, but the judgements levied on them were free of the kind of paternalism that somewhat softened those made about Native Americans.

Fanny Kemble, an English actress who recorded her observations during a visit to the American South in her *Journal of a Residence on a Georgia Plantation in 1838–1839*, gives a scathing description of the white settlers who lived in the barrens and swamps:

> These wretched creatures will not, for they are whites (and labor belongs to blacks and slaves alone here), labor for their own subsistence ... Their food is chiefly supplied by shooting the wild-fowl and venison, and stealing from the cultivated patches of the plantations nearest at hand. Their clothes hang about them in filthy tatters, and the combined squalor and fierceness of their appearance is really frightful.[43]

George Catlin,
Osceola, 1837,
lithograph.

White settlers in the swamps were subject to even harsher critiques by those who dwelt near them. William Byrd II, the archetypal gentleman who charted the dividing line through the heart of the Great Dismal Swamp, blasted the 'lubbers' – his term for the indolent, slothful whites who dwelt in the swamp and subsisted off its easily got bounty. Byrd, profoundly defensive as a gentleman living in a place regarded by many in more 'civilized' places as an unredeemed backwater, sees the 'lubbers' who live at the edges of the swamp as contemporary lotus-eaters, lulled into laziness by the swamp's poisonous bounty. As Byrd recounts in his *History of the Dividing Line Betwixt Virginia and North Carolina* (1728),

View of marsh at
Richmond Hill
Plantation, Georgia.
From the Historic
American Landscapes
Survey, *c.* 1930.

drones . . . are but too common, alas, in that part of the world. Though, in truth, the distemper of laziness seizes the men

oftener than the women. These last spin, weave, and knit,
all with their own hands, while their husbands, depending
on the bounty of the climate, are slothful in everything
but the getting of children, and in that only instance make
themselves useful members of an infant colony.[44]

Unmanly sloth infects the men thanks to the swamp's easily had
gifts. As we have seen, swamps were to be tamed, drained, conquered – to live comfortably at their edge meant that one was
either a primitive or, worse, a sluggard.

The often-repeated accusations of sloth among swamp-
dwellers were not tied simply to the ease of living off the swamp's
bounty. The noted naturalist Georges-Louis Leclerc, Comte de
Buffon, claimed that because America was 'filled with moist,
unhealthy vapors, where "everything languishes, decays, stifles"',[45]
animals in the New World were small and stunted compared to
those in the Old, and that humans living in such a climate could
not help but be lazy and shiftless. Jack Kirby attributes the
famous laziness of swamp folk to malaria, which 'weakened and
dispirited, imposing limits on human effort and explanation. It
seems reasonable to posit malaria, then, as a principal cause of
hinterlanders' vaunted ambitionlessness.'[46] Because the poor did
not have the luxury of seeking out more hospitable climes in
seasons when malaria peaked, sloth became all the more
emphatically linked with social class.

Regarded by most as crooks, fugitives and ne'er-do-wells,
swamp-dwellers provided comic fodder, as well, for such writers
as the humorist Henry Clay Lewis. In his *Odd Leaves from the
Life of a Louisiana Swamp Doctor* (1843), Lewis writes in the
tradition of the southwest humorists, rendering the quasi-
autobiographical foibles of Madison Tensas, his narrator, as he
calls on patients in the Louisiana swamps. Tensas's cases include
an elderly woman who has run out of whisky and is suffering
from delirium tremens, and whom he must supply with more
liquor without letting on to her solicitous friends that he is doing
so, and a puffed-up Southern gentleman who is determined that
he and his family, disease-ridden as they are, shall be known to

suffer from only the most aristocratic of ailments. Throughout, Lewis skewers the pretensions of antebellum Southern society as represented and belied by these comical, swamp-dwelling hinterlanders. Underlying Lewis's humour, though, is a sense of grotesquerie and horror. As Edwin T. Arnold puts it, Lewis finds humour in the 'disturbed, the deformed, and the dispossessed, the physical and psychological "monsters" who inhabit the borderlands between solid earth and liquid swamp, and between the rational world and the world of madness'.[47] While swamp-dwellers could be sources of humour, their strangeness, their perceived sloth, their very distance from order and light and civilization, made them menacing, if fascinating, creatures.

The dangers posed by such people in the antebellum South were not only moral, but practical. Jack Kirby describes the swamps as pockets of resistance to mainstream Southern culture and society for those outside the plantation order:

> This same wetland environment rendered free humans freer to resist both bourgeois society and the agronomic reformers ... In and near the Great Dismal, especially, woods-burning and hog-running country folk might live their 'careless' lives ... and still raise cash at will, on the periphery of the world's market order.[48]

Describing the customers of an erstwhile general store located on the outskirts of the Great Dismal Swamp in the nineteenth century, Kirby underscores the swamp-dwellers' anti-consumerism: they were

> mostly plain swampers, modest in their commercial require-ments, not yet 'consumers' in the modern sense. Perhaps such folks were without consumerist ambition, trading what they found or made from nature only to acquire essentials that were unavailable locally.[49]

Trading skins of small game for practical necessities, these swampers stood in contrast to the values that were supposed to animate and drive culture. The problem such an example posed for plantation owners lay in the example it set for slaves. Frederick Law Olmsted tells of a conversation with a plantation owner complaining of the lazy ways of some swamp-dwelling Acadians near his farm:

> Mr R. described them as lazy vagabonds, doing but little work, and spending much time in shooting, fishing, and play ... Why did he dislike to have these poor people living near him? Because, he said, they demoralized his negroes. The slaves seeing them living in apparent comfort, without much property and without steady labor, could not help thinking that it was not necessary for men to work so hard as they themselves were obliged to; that if they were free they would not need to work.[50]

Henry Louis Stephens, 'In the Swamp', from *Album Varieties No. 3: The Slave in 1863* (1863).

To fugitive slaves, the swamps were havens – fraught with dangers, certainly, but also providing a measure of shelter and safety from discovery. Swamps were natural refuges – dark, difficult to navigate, largely uncharted and forbidding. Nat Turner and his band of insurrectionists hid out in the Virginia swamps, as would Osceola and the Seminoles in Florida. Harriet Jacobs, in her narrative *Incidents in the Life of a Slave Girl*, describes an escape through the menacingly named Snaky Swamp. Beset by the swamp's terrors, swarmed by mosquitoes and surrounded by snakes that she and her companions have to keep at bay with sticks, Jacobs prefers a natural hell to

a man-made one: 'even those large, venomous snakes were less
dreadful to my imagination than the white men in that community called civilized.'[51] While Jacobs's time in the swamps was
short, popular legends peopled the swamps with untold numbers
of escaped slaves, grown savage and feral by their distance from
civilization. As we have seen, these legends contributed to
swamp monster myths that persist to this day.

Even for slaves who did not attempt to run away permanently, the swamps offered opportunities. Frederick Law Olmsted,
in his *Journey in the Seaboard States*, recounts a conversation with
a plantation owner who tells him that often a slave,

> thinking he is worked too hard, or being angered by
> punishment or unkind treatment . . . takes to 'the swamp,'
> and comes back when he has a mind to . . . His owner, who,
> glad to find his property safe, and that it has not died in
> the Swamp, or gone to Canada, forgets to punish him, and
> immediately sends him for another year to a new master.[52]

Living off livestock they would take from surrounding farms,
slaves could live in the swamps indefinitely, according to Olmsted's
account.

Slaves who worked in the swamps represented an entirely
different situation. On the one hand, slaves who worked in the
swamps found in them a measure of liberty compared to their
usual work. Edmund Ruffin, the agricultural reformer and ada-
mant supporter of slavery, describes these slaves as relatively
privileged:

> they live plentifully, and are pleased with their employment
> – and the main objection to it with their masters . . . is
> that the laborers have too much leisure time, and of course
> spend it improperly . . . given to idleness, and by many to
> drunkenness.[53]

But for all the benefits and possibilities that swamps presented
for enslaved African Americans, they could also be hellish places

to work. George Washington's Great Dismal Company, discussed in another chapter, saw staggering numbers of enslaved workers die in the swamps, and no reliable records exist to show how many slaves who sought shelter in the swamps succumbed. The story of slaves in the swamp, then, is, like so many other swamp stories, one of contrasts and contradictions: of limited freedom, of compromised shelter, of opportunity and tragedy.

A uniquely swamp-identified culture, the Acadians of Louisiana have a fascinating history in terms of their relationship to the swamps and to their own cultural distinctiveness. French colonists who had settled in Acadia, a colony located in what is now coastal Canada and Maine, the Acadians were driven out by British colonists during *Le Grand dérangement* or the Great Expulsion, beginning in 1755. Many Acadians eventually settled in Louisiana. As Marjorie Esman explains in her article 'Tourism as Ethnic Preservation: The Cajuns of Louisiana', 'Cajuns created their own culture based on economic adaptations to Louisiana environmental conditions and strongly influenced by Catholicism. Until the middle of [the twentieth] century most Cajuns were isolated and poor, and had little opportunity or desire to mingle with others.'[54] The Cajuns learned to make the most of the swamps of southern Louisiana, fishing, hunting and trapping, developing a distinctive culture closely associated with the Louisiana wetlands.

Though in some ways they fit the pattern of swamp peoples as refugees and outcasts who live off the land, the image of Acadians as an undifferentiated culture of happy-go-lucky people of the swamps is reductive. Carl Brasseaux, one of the most distinguished historians of the Cajun people, explains that

outsiders have consistently viewed the Acadians as a monolithic group of honest but ignorant and desperately poor fishermen and trappers, clinging tenaciously to an ancient way of life in the isolation of Louisiana's swamps and coastal marshes. Indeed, some writers have suggested that Acadiana has remained relatively unchanged since the time of the Acadian migration to Louisiana.[55]

Through the course of his book *Acadian to Cajun: The Transformation of a People, 1803–1877*, Brasseaux depicts an idyllic non-materialistic culture of farmers and labourers in which 'the poorest predispersal Acadian considered himself no less worthy than his wealthiest neighbor';[56] this culture gradually changed, after its dispersal from Nova Scotia and settling in Louisiana, to accept American materialist values and class stratification. While the Acadians initially hunted, trapped and fished among the Louisiana swamps, their culture was by no means stagnant. Acadians began to split among themselves. Many emulated the Creole planters around them, adopting slavery and transforming the landscape. They cleared swampland and planted cotton and rice, leaving hunting, fishing and trapping behind and emulating first Creoles, then Anglo-American farmers. As Brasseaux explains, though,

> many Acadians . . . rejected this materialistic mentality and sought to perpetuate their traditional life-style in the relative isolation of the lower Lafourche Basin and in the vast prairies of southwestern Louisiana. Between these polar extremes lay a majority of the descendents of the Acadian exiles, who found themselves torn increasingly between the self-sufficiency of the past and the materialism of the present.[57]

There emerged among the Acadians a split between the planter class and the *petits habitants*, a yeoman class who typically 'rejected American ideals, preferring instead their fathers' precapitalistic values and folkways'.[58] These were the swamp-dwellers, those who resisted the dominant culture in favour of a simpler, more traditional way of life. They 'resented the self-proclaimed superiority and sanctimony of their social "betters," while members of the social aristocracy were chagrined by Acadian "stubbornness" and "impudence"'.[59] An insular people who distrusted education and assimilation, the *petits habitants* represent the Cajuns that most outsiders – and even most contemporary Acadians – imagine when they think of Cajun folk culture.

The celebration of that folk culture is a relatively recent phe-   Sunrise with canoeists
nomenon. While most regard Cajun culture as happy-go-lucky,   near Bayou Benoit.
characterized by dance, good food and general indulgence and
festivity, as Eric Wiley puts it,

> The image of Cajuns as a fun-loving people is deceptive.
> Veiled behind it are a people still reeling from a government-
> sponsored assault on their culture and language, which arose
> from a national effort, beginning in the 1920s, to bring
> subcultures into greater conformity with the Anglo
> American mainstream.[60]

Acadian children were forbidden to speak French in public
schools beginning in the mid-1920s, and the prohibition was
often enforced through corporal punishment. 'Over time,' Wiley
explains, 'Cajuns came to feel ashamed of their language and
heritage.'[61] For much of their history, Acadians were essentially
shamed out of their own culture, derided for language differences
and compelled to assimilate.

A shift towards cultural reclamation and pride came with
the emerging tourism industry. Because, as Esman explains,
'Tourism and ethnic pride emerged at roughly the same time
in Louisiana, during the 1960s,' many contemporary Cajuns are
'tourists within their own culture'.[62] Alienated from traditional
Cajun life by decades of acculturation and modernization,

> Contemporary Cajuns are interested in their traditions
> and heritage, but they have little desire to adhere to them,
> having worked so hard to surpass them. They want to
> preserve a separate identity but also to participate in
> mainstream American culture. For most Cajuns today,
> especially those of middle age and below, expressions
> of ethnic pride largely substitute for the actual cultural
> distinctiveness that they proclaim.[63]

Any culture defined by rustic simplicity, by age-old tradition and
resistance to modernization, is perhaps inevitably doomed to be

consigned to the past. For swamp cultures, linked to endangered spaces, vestiges of wildness in a world where wildness is no longer dangerous but must be carefully protected, this holds particularly true.

Like the Acadians, some contemporary heirs of swamp- or bog-dwelling cultures look to the landscape to find traces of a fading cultural heritage and history. The great Irish poet Seamus Heaney finds in the bogs of Ireland, with their peculiar

preserving properties, a locus of cultural memory. As he puts it in his famous essay 'Feeling into Words',

> I began to get an idea of bog as the memory of the land-scape, or as a landscape that remembered everything that happened in and to it . . . I had a tentative unrealized need to make a congruence between memory and bogland and, for the want of a better word, our national consciousness.[64]

Heaney's poem 'Bogland' emphasizes the bogs' ability to contain and preserve the distant past, unspoiled, unadulterated. Beginning with the image of sunken butter recovered from the bog depths 'salty and white' after more than a century, the poem likens peeling away layers of bog to peeling away layers of cultural memory: 'Every layer they strip / Seems camped on before. The wet centre is bottomless.' Sometimes, the memories Heaney plumbs from the bogs in his poems are meditations on violence and cruelty. In 'Punishment', Heaney uses the discovery of the well-preserved, centuries-dead body of a young woman in a peat bog as an occasion to consider the burdens of inherited and contemporary cultural and personal guilt. The poem is inspired by the discovery of a teenaged girl in a bog in Germany in 1951. The girl's body, preserved since the first century AD, showed signs of her having been punished for adultery: shaving the head and weighting the body with stones were elements in punitive ritual. Heaney's speaker meditates on the body – 'I can see her drowned / body in the bog, / the weighing stone, / the floating rods and boughs / . . . her shaved head / like a stubble of black corn' (ll.9–12, 17–18). The speaker is moved by the poor, under-nourished adulteress, but ultimately recognizes his own complicity, recalling his silence in the face of the IRA's mistreatment of women in Ireland who were perceived to be fraternizing with the British troops. His guilt is testament to the timeless evil disinterred from the bog: 'My poor scapegoat, / I almost love you / but would have cast, I know, / the stones of silence' (ll.28–31).

Fisherman trawling for his catch in a swamp.

Heaney's speaker, despite his civilized, contemporary perspective, feels a primal link to the 'tribal, intimate revenge' here, and finds context and understanding in the distant past, in the fossils, the bodies, the memories preserved in the bogs. The poem is one of several, including 'The Tollund Man', 'Bog Queen' and 'The Grabaulle Man', in which Heaney meditates on the remains of the dead exhumed from the bogs of Europe, and on Irish history. Bottomless preservers of Ireland's past, the bogs represent, for Heaney, 'an answering Irish myth' in response to the American myth of the frontier.[65]

In the United States, few if any true swamp-dwellers remain. The Seminole did not, of course, simply die out, and their twentieth-century history is, again, linked fundamentally to the landscape. The drive to drain and develop the Everglades had drastic effects on the Seminole's subsistence within the swamps. As the official website of the Florida Seminole tribe explains, by the 1920s 'the wilderness no longer offered salvation; many [Seminoles] lived as tenants on lands or farms where they worked or as spectacles in the many tiny tourist attractions sprouting up across tourist South Florida.'[66] As swamps were drained, developed, marketed as tourist attractions and ultimately preserved by federal law, so were the Seminole impoverished, assimilated, put on display and ultimately apportioned an 80,000-acre (32,000-ha) reservation in 1938. Once there, they built a thriving economy, exempt from taxes, building enterprises including tobacco shops and eventually casinos. As the website reports,

> Today, most Tribal members are afforded modern housing
> and health care. The Seminole Tribe spends over \$1 million
> each year on education, alone, including grants-in-aid
> to promising Tribal college students . . . Dozens of new
> enterprises, operated by Tribal members, are supported
> by both the Tribal council and board.[67]

The Seminole, the unconquered, persist and thrive, but bear little resemblance to the swamp-dwelling 'savages' of a century ago.

If the official website of the Seminole tribe is any indication, any vision of alternative, pre- or anti-capitalist swamp-dwellers

Swamp hut in
the southern U.S.

has been supplanted by savvy entrepreneurship and marketing. Alongside links to capsule summaries of Seminole history and culture are links to swamp tours and RV resorts, as well as postings of job opportunities for the Hard Rock Hotel and Casino locations in Hollywood and Tampa. The Seminole have not died out, have not vanished, as wistful, nostalgic fantasy predicted over a century ago; instead, they have shed the trappings of the swamp-dweller: no longer menacing savages or anti-modern emblems of simplicity, the contemporary Seminole are, according to their website, a self-governed tribe of tech-savvy businesspeople, major employers and casino moguls. While the site emphasizes on its homepage that the Seminole are 'the only tribe in America who never signed a peace treaty',[68] the contemporary Seminole here seem a far cry from the defiant rebels who guarded their separateness in the depths of the Florida swamps. Here, custodians and marketers of cultural traditions, they are fully fledged participants in the local and global economies.

Oil well in a cypress swamp, Louisiana, 1930s.

For Acadians, embracing modernity has unfortunately meant sacrificing the landscape that contributes so heavily to their cultural distinctiveness. Many work in the oil industry, which continues to pollute and endanger the marshes and swamps of the Gulf Coast. While swamp recreation is still a major part of Cajun life, most experience it as a weekend getaway, in hunting and fishing camps where they escape from twenty-first-century life. As Esman explains,

> for urban or acculturated Cajuns [hunting and fishing] have been reduced to the level of hobby or sport. Maintaining a 'camp' in the swamp or marsh, used for weekend hunting and fishing, is common and desirable. Cajuns retain a sentimental attachment to the swamp environment and those who are familiar with it take pride in that knowledge.[69]

Esman's observations pertain to 'urban or acculturated Cajuns' – a large and growing majority in the twenty-first century, to be sure. But what of the few who still live largely as their ancestors did, by hunting, fishing and trapping in the Louisiana swamps? Some still remain; most pursue this way of life by choice rather than necessity. While the swamps remain a major part of Acadian culture and life, the combination of simplicity and exoticism inherent in living among them has attracted its own postmodern industry. The Arts and Entertainment network, or A&E, has found a reality television hit with *Swamp People*, a show that focuses on alligator hunters in the Louisiana swamps, who, often in competition with one another, drag great scaled leviathans out of the waters for camera crews and curious audiences. Their alienness is underscored at every turn – by the show's title, which brings to mind B-movie horrors of the 1950s; by the subtitles that accompany the thickly accented English spoken by many of the show's regulars; by the occasional family gatherings at the stilt-propped houses of Pierre Part, a tiny swamp community reachable only by boat, where featured hunters extol quaint, 'authentic' values of family togetherness and tradition. Amid the obvious reality television artifice, the 'Swamp People' stage their cultural show, part extreme wildlife circus, part game show, part family drama, doled out in 22-minute increments, between commercials and advert breaks.

2 Swamp as Quagmire: Obstacle, Trial and Problem

Away to the Dismal Swamp he speeds –
His path was rugged and sore,
Through tangled juniper, beds of reeds,
Through many a fen where the serpent feeds,
And man never trod before.
Thomas Moore (1779–1852), 'A Ballad: The Lake of the Dismal Swamp'

It is fitting that 'swamp' is a verb as well as a noun. As practical obstacles, swamps have been nearly unparalleled among geographic features. Mountain, sea, desert – none holds quite the same power as swamp for sheer, stubborn intractability. Menaces or nuisances, horrors or hassles, proving grounds or financial sinkholes, swamps have a long, rich and varied history as quasi-active antagonists to human endeavour, whether collective or individual, programmatic or personal. To 'swamp', as the *Oxford English Dictionary* defines it, is 'to overwhelm with difficulties', 'to ruin financially'; the *World English Dictionary* adds 'to render helpless'. In its adjective form, 'swamp' is similarly negative; to become 'swamped' has similar connotations – when one is swamped, one is buried under unmanageable tasks, weighted down by work, overcome by seemingly insurmountable labours. Such terms as 'mired' and 'bogged down' similarly link swampy landscapes to onerous burdens. While the reputations of swamps as a whole have undergone considerable rehabilitation over the past couple of centuries, the sense of swamps as irredeemable morasses and practical hazards retains considerable power.

The practical problems presented by swamps and wetlands, however, have shifted dramatically since the turn of the twentieth century. As the essential benefits of swamps have become known, and as technological advances have replaced the impervious, wild swamps with shrinking, threatened wetlands, the question of how to cope with the swamps has led to new quandaries and quagmires. This chapter explores the evolution of swamp as practical obstacle or problem, and the historical and cultural shifts that have driven that evolution. In a sense, the swamp's changing status as practical obstacle drives the attitudes and representations discussed throughout this book.

Foul, wild, intractable swamps make for a rhetorically powerful image. Harriet Beecher Stowe, the famed abolitionist and author of *Uncle Tom's Cabin*, invokes it in her novel *Dred* as a symbolic emblem of Southern moral depravity:

> The wild, dreary belt of a swamp-land which girds in those states scathed by the fires of despotism is an apt emblem, in its rampant and we might say delirious exuberance of vegetation, of that darkly struggling, wildly vegetating swamp of human souls, cut off, like it, from the usages and improvements of cultivated life.[1]

Speaking at Capon Springs, Virginia, in 1851, to a crowd doubtless quite familiar with swamps, Daniel Webster used it to characterize the prospect of secession:

> Those who love the Union ardently, and who mean to defend it gallantly, are happy, cheerful, with bright and buoyant hopes for the future . . . but secession and disunion are a region of gloom, and morass, and swamp; no cheerful breezes fan it, no spirit of health visits it; it is all malaria. It is all fever and ague . . . Nothing beautiful or useful grows in it; the traveler through it breathes miasma, and treads among all things unwholesome and loathsome . . . For one, I have no desire to breathe such an air, or to have such footing for my walks.[2]

Even as, as we shall see, notions about swamps breeding disease and pestilence by their very air have been corrected by science, the swamp retains its rhetorical force. While Washington, DC, was built on what is more accurately described as a tidal plain than a swamp, the metaphorical link between the centre of United States government and a quagmire of politics, special interests and potential corruption has proven compelling enough that the *Chicago Tribune*'s Washington Bureau blog takes 'The Swamp' as its name. As the *Tribune*'s website explains,

> The Swamp, born in 2006, is a triple entendre: The city was built on a swamp. Capitol Hill press conferences occur in a spot known as the Senate Swamp. And Washington often can seem to be a morass of partisan politics, political intrigue and complex legislation and policy. Our goal is to help readers navigate the Swamp.[3]

The Great Dismal Swamp is a marshy area in the Coastal Plain region of southeastern Virginia and northeastern North Carolina.

To 'drain the swamp' is a common metaphor for taking the necessary steps to clear out an entire corrupt or unmanageable system, be it political, criminal or otherwise. The term has even been used recently in the context of global terrorism. U.S.

Lucius Junius
Moderatus Columella,
16th-century portrait.

Deputy Secretary of Defense Paul Wolfowitz, seeking NATO help in capturing terrorists in 2001, used memorable imagery in depicting u.s. strategy: 'While we'll try to find every snake in the swamp, the essence of the strategy is draining the swamp.'⁴ More recently, during his 2016 presidential campaign, Donald Trump made 'draining the swamp' in Washington, DC, a central promise, targeting an inchoate sense of corruption and cronyism underlying u.s. government. Long after physical swamps have been recognized as fragile, endangered wetlands, rhetoric sustains their image as menacing morasses, havens for all manner of evil.

Disdain for swamps as practical hazards and obstacles to civilized cultivation dates back millennia. The Roman agricultural writer Columella, writing in the first century AD, blames marshlands for a vivid catalogue of hazards and foulness. Marshland, Columella reports, throws off a baneful stench in hot weather and breeds insects armed with annoying stings, which attack us in dense swarms; then too it sends forth plagues of swimming and crawling things deprived of their winter moisture and infected with poison by the mud and decaying filth, from which are often contracted mysterious diseases whose causes are even beyond the understanding of physicians.[5]

Edward Lear, 'The Pontine Marshes from above Terracina', *c*. 1880, pen in brown ink. Caption reads 'Stretched wild and wide the waste enormous marsh.'

The drinking of swamp water, Columella believed, was potentially deadly unless the water was purified: 'Worst of all is swamp-water, which creeps along with sluggish flow; and water that always remains stagnant in a swamp is laden with death.' The swamps' foulness could be overcome, though, by winter rains, as 'water from the heavens is known to be most healthful, as it even washes away the pollution of poisonous water.'[6] The antagonism here between the purifying waters of heaven and the virulent, deathly poison of the swamp leaves little doubt as to the latter's moral implications.

Nº 8

" Stretched wild and wide the waste enormous marsh. "

The Pontine Marshes from above Terracina.

Often virtually personified by their real and perceived dangers, swamps were imputed the ability not only to act – to swamp – but to breathe. One of the oldest and most persistent beliefs about the dangers swamps posed to humans is the idea that swamps' exhalations caused disease. Indeed, malaria, which comes from the Italian for 'bad air', was for a long time a catch-all term for any disease or ailment thought to result from inhaling swamp miasma:

> to explain the occurrence of periodic fevers, eighteenth century Italian writers suggested that effluvia or exhalations from the marshes or from within the earth caused the fevers. This atmospheric poison or miasma came to be called *mal'aria* . . . During most of the 19th Century, malaria was not used as the name of a disease; it referred to a noxious material, presumably gaseous, which emanated from swamps or from rotted vegetation or decaying animal carcasses.[7]

Although Alphonse Laveran traced the true cause of malaria to the parasite *Plasmodium* in 1880, the idea that swamps generated and essentially breathed disease was pervasive, persisting among the general public well after Laveran's discovery.

Dispatches from the early American colonies are full of critical and disparaging depictions of swamps, and most often

Advertisement for Ayer's Ague Cure malaria medicine, *c.* 19th century.

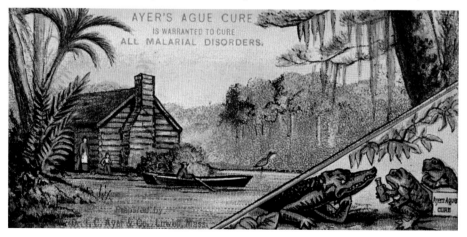

reflect the common belief that their exhalations bred disease. Stagnant water meant all manner of disease, including typhoid and yellow fevers in addition to deadly malaria. The association was so powerful among the colonies that it rendered 'wetlands and death inseparable'.[8] Albert Cowdrey, in *This Land, This South*, explains that swamps were often scapegoated for diseases that originated elsewhere (that had been imported with the slave trade and thrived in swampy regions):

> Yaws showed up among slaves in Jamaica and spread widely in the South among whites, blacks, and Indians. In one region of North Carolina near the Dismal Swamp, its lesions were so common in the early eighteenth century that William Byrd II jocularly reported a supposed motion in the colonial assembly to deny preferment to any man who had a nose.[9]

The diseases that ravaged the colonists, slaves and Native Americans during the colonial era were generally imported, either from Europe with the colonists themselves, or, later, from Africa and the Caribbean with slaves, then spread by the ubiquitous mosquitoes, which thrived in hot, damp swamp environs.

These explanations, of course, would be discovered later – most records of the time followed the age-old pattern of blaming swamps and marshes themselves for illness and pestilence. Captain Nathaniel Butler, writing in 1622, penned a devastating indictment of the Virginia colony as he found it, entitled 'The Vnmasked face of our Colony in Virginia as it was in the Winter of the yeare 1622.' Butler's account renders a barren colony situated in a diseased landscape: 'I found the Plantations generally seated vppon meer Salt Marrishes full of infectious Boggs and muddy Creeks and Lakes, and therby subjected to . . . inconveniences and diseases.'[10] Bernard Romans, in his *Concise Natural History of East and West Florida* (1775), deals at great length with what he calls 'the universally dreaded, though chimerical unhealthiness of this climate', which manifests in part in 'unhealthy swamp soil'. Coastal marshes and swamps

alike are putrid incubators of pollution and disease in Romans's account. St John's, situated among coastal marshes, is 'subject to thick nasty fogs of all kinds . . . these marshes a little before rain emit a most horrid, and to me a suffocating stench'. Though none of the residents seem to notice the smells, or speak of any unhealthy effects, Romans remains convinced that 'where this kind of marshes are situate in brackish water, the situation is beyond a doubt very unhealthy, the wan complexion and miserable mien of the generality of the inhabitants of such districts too plainly evinces it.' Romans also laments the 'uncommon swarms of flies, gnats, and other insects which attend putrid air', claiming that as these swarms of vermin die and decompose, 'their effluvia, even of those that are imperceptible to the naked eye, arising or exhaling from ponds, marshes, swamps, &c. must spread a great quantity of noxious vapours through the atmosphere, and

Alligator (*Alligator mississippiensis*) submerged in a swamp.

consequently corrupt the air, and spread disease throughout their vicinity.' Romans hints at a developing pattern in discussing the health effects of swamp environments on their dwellers by associating their harm with moral turpitude: 'let me advise every new comer, particularly a person of a gross habit of body, to be careful of his constitution, a wine-bibber or rum guzzler, with such a plethoric habit, can hardly avoid falling a prey to this bad air.' I will discuss this attitude more fully in another chapter, but here it is significant to note the moral valence imposed upon the swamps and their dangers, long a key element in their demonization.[11]

Disease was far from the only danger posed by swamps, bogs and marshes. Many of the dangers came from their inhabitants, ranging from stinging and biting insects, to poisonous snakes, to the alligator, who has captured the popular imagination as saurian sovereign of the dark and dangerous swamp domain.

Many of the hazards in swamps and bogs, though, come not from their inhabitants, but from the landscape or earth itself. The legend of the bottomless bog, into which the unwary traveller might sink forever without a trace, derives from this feature; indeed, archaeological evidence bears out that this was a genuine danger: 'the hands of a dead man recovered from Velne bog in Germany held tufts of heather, which he probably had grabbed in a vain attempt to pull himself out of the bog . . . Other bodies have been found clutching sticks.'[12]

Another significant bog hazard is the bog burst, which occurs when liquefying peat below the bog's surface causes great chunks of peat to shift dramatically, with sometimes catastrophic results:

> bog bursts present very real hazards to human beings and their possessions. Bog bursts have been recorded for centuries in Ireland, beginning with the 1697 Kapanihane bog burst in County Limerick. In 1708 another bog burst occurred in the same county, burying houses and people under 20 feet [6 m] of liquid peat.[13]

John William Orr,
'De Soto Preparing to
Cross Long Swamp',
in Lambert A. Wilmer,
*The Life, Travels and
Adventures of Ferdinand
de Soto, Discoverer of the
Mississippi* (1858).

DE SOTO PREPARING TO CROSS LONG SWAMP.

These bog hazards are often imported, however inaccurately, to swamps in the popular imagination. The swamps of film and fiction are full of quicksand, a phenomenon whose ubiquity and danger has been greatly exaggerated in popular culture. Fears of disappearing into the swamps never to be seen again are essential to the swamps' mystique as mysterious, unpredictable and, above all, dangerous places.

For all the misconceptions and exaggerations of swamp-related dangers, for most of history they have been undeniable obstacles to human endeavour – nemeses to some of history's most eminent figures. Some of the most distinguished and powerful figures in history have tried and failed, for various reasons, to drain swamps. Plutarch, in his life of Julius Caesar, reveals that the great emperor proposed 'to convert marshes about Pomentinum and Setia into a plain which many thousands of men could cultivate'.[14] As it turned out, Caesar was one of a long series of rulers, clergy and statesmen who would propose draining or attempt to drain the malaria-ridden marshes of Latium; centuries later, the Italian leader who would ultimately succeed was Mussolini, whose engineers drained and filled the marshes in 1939.

As mysterious, dangerous regions defined by their distance from civilization's control, swamps have long been depicted as archetypal terra incognita to be explored by gentlemen brave and bold enough to dare enter their depths. In the swamps of the American South, zones of pronounced conflict between European gentlemen and wild, untamed and impertinent nature, epic conflicts played out between man and nature, between progress and indomitable wilderness. European explorers since the earliest days of the colonial era set out on heroic errands through the swamps and marshes of the New World, seeking to tame, conquer and colonize.

William Byrd II's errand into the Great Dismal Swamp stands as a kind of symbolic, touchstone encounter between gentleman and swamp. Byrd was, since early life, eminently self-conscious about his status as a gentleman. As his biographer Kenneth Lockridge explains, Byrd strove throughout his life to fashion himself as a gentleman, following the values of 'moderation, temperance, and self-control',[15] and rejecting or mastering passion. Byrd's diary of his daily activities reveals his single-minded quest to live up to a gentlemanly ideal: 'Plainly, what the diary describes are the expected behaviors of an eighteenth

Wetland bogs in the USA.

Hans Hysing,
William Byrd II,
c. 1724, oil on canvas.

century gentleman repeated and obsessively reviewed.'[16] Byrd describes habitually rising at 4 a.m. to read verses in Hebrew and Homer's *Odyssey*. Eminently conscious of the potential judgement of others, particularly that of European gentry, who tended to regard the New World colonies in general as an uncultivated backwater and its inhabitants accordingly, Byrd carefully depicts himself in his more public writings as an eminent gentleman mastering nature within and without.

In *A History of the Dividing Line*, Byrd recounts his efforts to demarcate a line through the Great Dismal Swamp separating North Carolina and Virginia. Popular and widely read in Byrd's time, the work distils and exemplifies many fundamental assumptions and attitudes towards the swamp whose influence would persist and resonate for centuries. Byrd presents himself as a questing knight, willing to risk life and limb in the performance of his duty: as Lockridge describes it, he presents himself as 'a gentleman who was representative of a class of men which alone could direct the running of coherent lines through an incoherent wilderness'.[17]

Byrd characterizes the swamp as a fitting adversary for his questing knight: it is utterly unknown, he claims, even to those who live at its edges:

> It is hardly credible how little the bordering inhabitants were acquainted with this mighty swamp, notwithstanding they had lived their whole lives within smell of it . . . We saw plainly that there was no intelligence of this terra incognita to be got, but from our own experience.[18]

Not only is it mysterious, but it is utterly vile. Conscious always of lines and divisions between savage and civilized, nature and culture, Byrd repeatedly asserts that the swamp is more suited to vermin and reptiles than to human life. He acknowledges occasionally that the swamp holds some rare and strange beauties, but such features are to him temptations to moral failure, breeding sloth, an unforgivable sin, in those who succumb. To Byrd, the swamp's beauty is antithetical to life, and the environment, as his task progresses, becomes imbued with spiritual as well as practical menace:

> Doubtless, the eternal shade that broods over this mighty bog, and hinders the sunbeams from blessing the ground, makes it an uncomfortable habitation for anything that has life . . . It had one beauty, however, that delighted the eye, though at the expense of all the other senses: the moisture of the soil preserves a continual verdure, and makes every plant an evergreen, but at the same time the foul damps ascend without ceasing, corrupt the air, and render it unfit for respiration. Not even a turkey buzzard will venture to fly

over it, no more than the Italian vultures will over the filthy lake Avernus, or the birds in the Holy Land, over the salt sea, where Sodom and Gomorrah formerly stood.[19]

As vivid and religiously tinged as Byrd's judgements of the landscape are, they are best understood as, in part, products of his own project of self-representation. The more menacing to both body and spirit the mysterious swamp is, the more Byrd fits the figure of the knight errant, questing through a physical and spiritual waste.

Though his mission was successful and all of his men survived, Byrd's ultimate assessment of and prescription for the swamp was clear:

> The exhalations that continually rise from this vast body of mire and nastiness infect the air for many miles round, and render it very unwholesome for the bordering inhabitants. It makes them liable to agues, pleurisies, and many other distempers, that kill abundance of people, and make the rest look no better than ghosts. It would require a great sum of money to drain it, but the public treasure could not be better bestowed, than to preserve the lives of his majesty's liege people, and at the same time render so great a tract of swamp very profitable.[20]

Published almost a century after Byrd's death in 1841, his writings were widely popular; they both provided a glimpse into colonial attitudes towards the swamps and proved influential in shaping antebellum attitudes towards Southern swamplands.

More corporate ventures brought prominent gentlemen into conflict with the swamps, as well. In a fascinating historical episode, George Washington, who would, of course, go on to become the first president of the United States, co-founded a company with the goal of draining the Great Dismal Swamp. The appropriately named Great Dismal Swamp Company conceived of its goals as a combination of public service, transforming useless swampland into arable, productive soil and private profit. Charles

Royster, in *The Fabulous History of the Great Dismal Swamp Company*, describes Washington's ambitious beliefs regarding the swamp's soil:

> Most soil along the road into North Carolina was sandy and poor. Yet Washington was sure that within the swamp all was black and fertile . . . Though local people thought it 'a low sunken Morass, not fit for any of the purposes of Agriculture,' Washington felt certain that it was 'excessive Rich.'[21]

Assisted by laws passed by the House of Burgesses that shielded the company from any litigation by property holders, since the work they did was in the public interest, the Great Dismal Swamp Company spent more than forty years sending armies of slaves into the Great Dismal with the goal of clearing, draining and rehabilitating the swamp into arable land in what was ultimately a protracted and dramatic failure. For all the loss of capital and face it represented to Washington and his fellow investors, the real toll of the Great Dismal Swamp Company's efforts fell upon the labourers:

> suffering, disease, and deaths among slaves working on the Jericho Canal appeared in stories told by people in Nansemond County for generations – stories of 'chain-gangs of slaves': 'they say the poor creatures died here in heaps from swamp fever. But that didn't make any difference to their owners. They was made to dig right into the heart of the swamp to get at the juniper trees.'[22]

The company dissolved without achieving its goal, consoled by the profits they realized on selling the timber from the seemingly limitless trees the labourers felled without making a significant dent in the indomitable Great Dismal.

Military conflicts, too, were shaped by swamps. General Francis Marion, also known as the Swamp Fox, became legendary in part due to his ability to use the swamp terrain to his advantage during the American Revolution. Marion, as an archetypal gentleman soldier who seemed to master the un-tamable swamps, inspired Southern literary heroes in the works of writers like William Gilmore Simms.

During the Civil War, swamps became both strategic assets and, more often, practical obstacles. Ulysses S. Grant reports having to cut a wide swathe of timber under the waters of the cypress swamp outside Vicksburg, calling it 'an undertaking of great magnitude'.[23] The rank and file were often more colourful in their condemnations of the swamp. A Union soldier named

J. Wells Champney, 'A Peep into the Great Dismal Swamp', from Edward King's *The Great South* (1875).

Gilbert Stuart, *George Washington, Carroll Portrait*, 1912.

Thomas Mann blames the Chickahominy Swamp for a massive malaria outbreak among the men: 'Of the 600 out of every 1000 who did take some part in the prolonged struggle, many were weakened by the hydra-headed forms that the Chickahominy malaria took so as to incapacitate in degrees varying from slight indisposition to an almost helpless state.'[24] Civil War soldiers' accounts of suffering through swamps are many, as they waded through miles of swamp, suffered the attentions of 'swamp angels', the name some used for the great mosquitoes who plagued them, and tried to tame the wilderness in what limited ways they could.

Perhaps the most horrific tale of swamps in wartime is the Battle of Ramree Island, a six-week conflict in early 1945 between

the British Indian Army and the Japanese Imperial Army on an island off the coast of Burma (Myanmar). As the British forces drove Japanese troops across miles of coastal mangrove and inland swamps, the soldiers began to fall prey to swamp-related afflictions, including mosquitoes and swamp-related disease. Most dramatic, though, are legendary but apocryphal accounts of Japanese soldiers being devoured by crocodiles in what the *Guinness World Records* has dubbed the worst animal disaster in world history. As the story goes, a thousand Japanese soldiers entered those swamps in the dark of night, and only twenty emerged the next day. The rest had been devoured by crocodiles or otherwise claimed by the swamp.[25]

The legacy of all this swamp-centred strife remains powerful in contemporary culture. Even as actual swamps have nearly

Swamp on the Appomattox River, Virginia, near Broadway Landing.

John Sartain, 'General Marion in his swamp encampment inviting a British officer to dinner', 1840, engraving. Francis Marion, also known as the Swamp Fox, gained fame as the rare gentleman military commander who could seemingly tame the swamps for his own strategic purposes.

been supplanted in the popular imagination with the very real consciousness of fragile, endangered wetlands, the trope of swamp as a space of aestheticized danger-fraught wildness, as obstacle in a daring hero's journey, survives and thrives in literature and film. As we have seen, with the closing of the frontier, the swamps remained as pockets of pure wildness within the largely tamed American landscape. Adventure tales of rugged heroes venturing into their depths became popular in the early twentieth century, a kind of rugged, masculine outgrowth of the swamp vogue of the late nineteenth century. Whites who ventured into the Everglades in search of the Seminole tended to '[dwell] upon the difficulties they overcame as they journeyed to the Indians' Everglade homes', and '[emphasize] dangers ranging from rattlesnakes, alligators, and panthers to quicksand'.[26]

A story in the *New York Times* in 1889 epitomizes the notion of the swamp journey as ripping adventure yarn. The piece, entitled 'The Florida Everglades: Explorations of a venturesome few on Lake Okefenokee – The Remnant of the Seminoles',[27] presents a swamp landscape that defies not just the interloping white man, but even its own denizens: 'There is no shore to the lake, but dense swamps of saw-grass and wild, half-submerged hummocks encircle the clear-water sheet on every side, on which neither reptile, animal, nor man has ever yet found a footing.' These swamps are not just unexplored, but actively hostile to outsiders. Floating plants, 'like the treacherous will-o'-the-wisp, deceive the solitary travelers who venture into that lonely region', clogging the river's mouth and leaving explorers disoriented and unable to find their way. A single barrel, left by prior surveyors, which rests atop a single towering cypress 'that stands far out into the lake like some solitary sentinel on duty', represents 'the only mark of man visible within a radius of nearly 100 miles'.

A. R. Waud, 'Bridge through Chickahominy Swamp', between 1860 and 1865.

Gigantic saltwater crocodile caught in the mangroves of the Sundarbans in India.

The profound otherness of the swamps is the article's central theme. The author describes the vast Everglades as a place 'over which no surveyor's chain has ever been stretched, and of which all knowledge is as conjectural as of the interior of the Dark Continent beyond the path of Stanley'. The language here is telling, reflecting a vision of swamp as African other, again defined by a Western white perspective.

'To penetrate the Everglades is a daring undertaking,' the writer tells us, 'and none but the cowboys of South Florida ever offer to guide strangers into the dismal waste.' Even these rugged frontiersmen refuse to venture too deep into the Everglades, beyond the paths they have created: 'Dangers from rattlesnakes and alligators, which are as numerous as hairs on your head and of gigantic size, they despise; but when it comes to facing starvation in a trackless wilderness they weaken.'

One of the most interesting aspects of the article, and of the swamp mythology it reflects, is the sense that, for all the reptilian dangers it holds, the untracked, uncharted swamp itself is the real peril.

The land in the glades is treacherous, and in places a sort of quicksand will suddenly sink beneath one's feet and land you in a living grave. The bottoms of many of the lagoons are also so soft that one must swim across them and not trust to wading. A heavy rain will make the most marvelous changes in the country. Small streams will assume gigantic proportions in one night and by morning you will find yourself shipwrecked, as it were, on a small island, cut off from all surrounding land by long stretches of uncertain water. If the rains continue . . . you find your island gradually

J. Wells Champney, 'Shooting at Alligators', from Edward King's *The Great South* (1875).

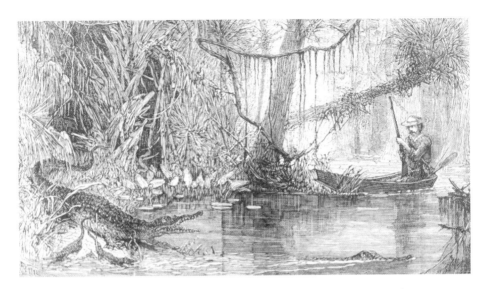

'A Sudden Turn in the Ocklawaha', from William Cullen Bryant, ed., *Picturesque America* (1874).

melting away from under your feet, and you strike out for another and larger one . . . Game of all description are keeping you company, and when you finally reach a larger and safer island bears, deer, wild cats, panthers, alligators, rattlers, and moccasin snakes are already there to give you a welcome.

While the presence of the swamp fauna might inspire fear, the author informs us that the flooding tends to render them docile and harmless, so long as one '[gives] them plenty of room'. Still, the flooding means that

> There is no escape from such prisons until the waters subside, and even then it requires many days for the most experienced woodsman to find his way back to civilization . . . If one does not starve to death in such a journey he is likely to be prostrated many days afterward with fever and exhaustion.

The archetype of the swamp adventurer retains considerable influence. Indeed, some of the most famous windows into the lives of swamp-dwelling peoples are provided by figures who are

rather mythologized themselves. Wilfred Thesiger, perhaps the most celebrated British chronicler of the culture of the Ma'dan, or Marsh Arabs of Iraq, is lionized by his disciple Gavin Young, who calls him 'the European who "discovered" the marshes'[28] and 'the greatest traveler of his time, and possibly any other'.[29] Depicting Thesiger as adventurer and man of action in the mould of Lawrence of Arabia, Young paints a vivid verbal portrait: 'the Marsh Arabs, who naturally admire physical prowess of any kind, were awed by Thesiger's ability to pursue the wild boars of the region on the saddle-less back of a temperamental Arab mare, with the reins in one hand and unerringly shoot the pigs dead, holding his Rigby .275 rifle in the other hand like a pistol.'[30] More recently, Nick Middleton tends to be cast as a world-travelling man of action: the *Sunday Telegraph*'s blurb on the cover of the paperback edition of *Surviving Extremes: Ice, Jungle, Sand and Swamp* calls him 'Geography's answer to Indiana Jones'.

Vivid evocations of swamp peril tend to underscore the heroism of the adventurer who navigates these treacherous, menacing domains. Indeed, passage through a forbidding swamp or marsh is a common element in the archetypal hero's journey in both ancient myth and in contemporary fantasy. In

J. Wells Champney, 'Some Tract of Hopelessly Irreclaimable, Grotesque Water Wildness', from Edward King, *The Great South* (1875).

J.R.R. Tolkien's classic *Lord of the Rings* trilogy, Frodo and Sam, guided by the sinister Gollum, must detour through the ominously named Dead Marshes en route to Mordor, their ultimate objective. Tolkien's Dead Marshes play to classic tropes of the menacing swamps, from their foul exhalations to their mixture of natural and supernatural menace; still, they are safer than the main roads, because even the dark lord's all-seeing eye cannot penetrate this mist-shrouded quagmire:

The fog and mist that rise from swamps can sometimes lend themselves to spectacular effects. Here, the Cocodrie swamp itself appears to blaze with fire.

It was dreary and wearisome. Cold clammy winter still held sway in this forsaken country. The only green was the scum of livid weed on the dark greasy surfaces of the sullen waters. Dead grasses and rotting reeds loomed up in the mists like ragged shadows of long-forgotten summers . . .

'Not a bird!' said Sam mournfully.

Westley and
Buttercup in
*The Princess
Bride* (1987).

'No, no birds,' said Gollum. 'Nice birds!' he licked his
teeth. 'No birds here. There are snakeses, wormses, things in
the pools. Lots of things, lots of nasty things. No birds,' he
ended sadly.[31]

The heroes' journey through the foul marshes includes
menacing ghost lights that threaten to lead them astray; the
faces of long-dead warriors appear beneath the grimy surface of
the water; once noble, now 'all foul, all rotting, all dead. A fell
light is in them.'[32] Once they successfully navigate the foul and
funereal marsh, they emerge 'slimed and fouled almost up to
their necks . . . [they] stank in one another's nostrils.'[33]

A tongue-in-cheek variation of the swamp as hero's obstacle
comes in William Goldman's comic, deconstructive fairy tale *The
Princess Bride*, the film version of which has become an enduring
cult classic. In it, the dashing hero Westley must lead his beloved
Princess Buttercup through the menacing Fire Swamp. Perhaps
inspired by the bog burst phenomenon, Goldman's Fire Swamp
is apt to burble momentarily before blasting a geyser of flame
straight up from the earth itself, in a dramatically exaggerated
twist on the idea of the personified swamp attacking interlopers
directly. Despite this danger and that of the R.U.S.s, or Rodents
of Unusual Size, which turn out to be giant, man-eating rats,
Westley leads his fair charge to safety on the other side of the
swamp.

Another memorable swamp vision from fantasy film comes
in the *The NeverEnding Story* of 1984. As Atreyu, the film's youth-
ful hero, seeks Morla, the wisest creature in the land of Fantasia,
he traverses locations with names like the Silver Mountains and
the Desert of Shattered Hopes. His journey takes him and his
horse, Artax, to the Swamps of Sadness, a metaphorical landscape that gradually draws anyone who lets sadness overtake
him into its murky depths. The swamp's sadness is inherent,
miasmatic, infectious; many a child was traumatized at the sight

The horse Artax sinks
into the swamp in
*The NeverEnding
Story* (1984).

of Artax, overcome by sadness, gradually sinking to his death
as a tearful Atreyu begged him to resist it and move on. Borne
down by hopelessness and pursued by swamp beasts, Atreyu

himself nearly succumbs to the swamp, until he is rescued by the Luck Dragon, Falcorr.

The swamp's appeal in fantasy storytelling extends to science fiction. In George Lucas's *Star Wars* trilogy, the hero, Luke Skywalker, journeys to the swamp planet of Dagobah to seek out his diminutive mentor, Yoda. There, in the heart of the swamp, he undergoes rigorous training that culminates in his lifting his spaceship from the swamp mire through sheer force of will, harnessing the semi-magical powers that Yoda has trained him to use. Passage through and escape from the mire is a trope that resonates in contemporary pop culture across multiple genres: any number of video games, for example, feature swamp levels or worlds, populated with various beasts and snares that must be navigated and overcome before the hero can progress.

While the vision of the forbidding, uncharted swamp is firmly enshrined in popular culture, the mid-nineteenth century

Georges-François Mugnier, 'Lumberjacks Logging Cypress Swamps', *c.* 1880–1920.

saw the beginning of its end in reality. Inevitably, technological progress eventually overcame the swamps' legendary indomitability. In some parts of the American South, flooding was a constant danger, and swamp drainage became an essential part of flood control. Drainage was both necessary for the public good and a potential boon to speculative investors. The necessity of draining flood-prone swamps led to the passage of the Swamp Land Act of 1849. This act essentially stated that wetlands that risked severe flooding could be designated as the property of individual states, rather than of the federal government. As such, they could be sold, with the proceeds used to improve drainage and build levees. The passage of this act made possible the swampland timber industry, which hastened the American swamps into a different kind of practical quagmire.

Once technology made the prescriptions of Byrd and his would-be swamp-clearing antecedents dating back to Caesar and Columella possible, clearance and drainage proceeded at a remarkable rate. In America, reclamation became not only an economic investment, but, to some extent, a matter of national pride. In his 1907 report titled *Swamp and Overflowed Lands in the United States*, J. O. Wright proclaimed that

> After considering what has been done to reclaim the marshes of Holland, two-fifths of which lie below the level of the sea, and the difficulties that have been overcome in draining the fens of England, it would be a reflection on the skill and intelligence of the American engineer to proclaim the drainage of our swamp lands impossible.[34]

Unfortunately, the practical intelligence and technological capability of engineers outstripped people's understanding of the swamps' various beneficial functions.

In the early twentieth-century American South, the once indomitable swamps came briefly to seem like inexhaustible resources and potential salvation from the crippling poverty that had beset the region since the Civil War. Describing southwest Louisiana in *The Great South*, Edward King saw 'more than

3,000,000 acres of land of almost inexhaustible fertility. The giant cypresses along the lakes and bayous are abundant enough to last for a century. Employment to hundreds of mills and thousands of workmen could readily be furnished, the lumber being easily floated down the innumerable bayous and along the lakes to market.'[35] King was right about the wealth of resources to be found in the timber-rich swamps; he was wrong, however, about how long those swamps could sustain an infusion of ambitious Northern industrialists. Developers, realizing the money to be made from Southern timber, poured into the Southern wetlands, cleared the old-growth trees, and then cleared out of the towns they had briefly revitalized, leaving the Southern swamps drastically deforested by the mid-1920s.

Southern swamps went from impenetrable wilderness to quickly exhausted sources of commercial timber with stunning rapidity. The swamps were beset by competing agendas: Nelson Manfred Blake, in his 1980 study *Land into Water, Water into Land*, explains that, while much Southern wetland loss can be blamed on timber barons – speculators who would come into a community, usually from the North, employ local workers for a while in clearing out the area's timber, and then depart, leaving

Felling trees for drainage, Miami, Florida, c. 1910–20.

'Dismal swamp
Canal', from *Harper's
New Monthly*
(May 1860).

the landscape denuded and the people unemployed – some
swamp-draining and -clearing initiatives were clearly intended
to benefit local economies. In the case of the initiative to drain
the Everglades in the early twentieth century,

> the politicians who authorized the digging of an ambitious
> system of canals and ditches saw themselves as champions of
> the people redeeming millions of acres of soil from wealthy
> monopolists and transforming these swampy tracts into an
> agricultural paradise for small farmers.[36]

Problems quickly began to emerge as swamp deforestation
and drainage continued their rapid progress. Hunters began to
notice fewer waterfowl in the skies; wildlife enthusiasts began to
realize the toll that the shrinking wetland habitats were having
on native animals; gradually, an infant conservation movement
began to coalesce. Other problems were more immediate and
practical. As Donald Hey and Nancy Philippi explain, 'the first
national awareness of the interrelatedness of the different features in a watershed came in the late nineteenth century with

the notion that the loss of forests, at the hands of the timber barons, was resulting in increasing floods and reduced water supply.'[37] In an age in which industry and progress were paramount, and practical conservation was a fledgling idea, the response to increased flooding was to construct more artificial flood controls, which had limited effectiveness. Gradually, over decades, engineers and resource planners came to realize the natural functions of the swamps that had been cleared – that much of the increased flooding was due to the fact that 'the wetlands that had once captured and then slowly released the heavy flood flows had been cut off from river channels by navigation and flood control projects.'[38] The natural functions that wetlands perform – slowing runoff, helping with flood control by soaking up water, filtering toxins and pollutants – have to be replaced by often expensive, artificial means, like treatment plants and dams, after those wetlands have been destroyed.[39]

Percy Viosca, a naturalist and freelance biologist who had witnessed the ravaging of the south Louisiana wetlands at first

Felling trees for drainage construction at New Market Creek Swamp, c. 1910.

hand, published one of the earliest conservationist arguments against wetland destruction in 1928 in an issue of the journal *Ecology*. As Viosca put it, 'Reclamation experts and real estate promoters have been killing the goose that laid the golden egg ... our future conservation policy should be restoration of those natural conditions best suited to an abundant marsh, swamp, and aquatic fauna.'[40] Despite the gradual emergence of perspectives like Viosca's, the quagmire would only deepen as economic depression in America led to the many paradoxical initiatives of the New Deal era. While agencies such as Roosevelt's Civilian Conservation Corps recognized the role of wooded wetlands in preventing and controlling floods, it seemed that for every wetland conservation initiative, there was a simultaneous initiative to drain or reclaim swamps in the name of economic amelioration. Thus different government agencies simultaneously sought to conserve wetlands and actively cleared and drained millions of acres of swamps and marshes. These contradictory efforts would continue until the passage of Executive Order 11990 under President Jimmy Carter, requiring all federal agencies to 'take action to minimize the destruction, loss, or degradation of wetlands'.[41]

Efforts to drain and 'reclaim' swamps around the world have met with varying results, and can provide interesting insight into the philosophies and priorities of leaders and nations. One of the best-known and most successful swamp-clearing campaigns of the early twentieth century came when Benito Mussolini, Italy's fascist dictator, undertook to drain the Pontine Marshes. Located approximately 50 km (30 mi.) south of Rome, these marshes had been the bane of civilized Italy for centuries. Flooded throughout most of the year, the marshes bred swarms of mosquitoes that were major contributors to the nation's outbreaks of malaria, which was so widespread in Italy that it had become known by Mussolini's time as the 'Italian National Disease'.[42] Mussolini undertook the task that had thwarted emperors and popes before him: to drain, clear and purify the marshes, to rid Italy of malaria, and to transform this lost land into fertile farmland and prosperous towns.

The project was part of Mussolini's *Bonifica Integrale* plan, and was driven by the same ideology that led him to want to 'purify' all aspects of Italian culture. He approached it as a military campaign, sending in tremendous numbers of workers, most of them military veterans. Working at a relentless pace, the workers created an enormous network of drainage canals, as well as six enormous pumps which suck as many as 9,500 gallons (43,000 l) per second of water out of the marshes and send it through those canals to the sea. Without the constant work of those pumps, the area would be marsh again within a week.[43]

The entire undertaking was framed for maximum propaganda value, garnering extensive coverage in newsreels and other media. Mussolini appeared in photo opportunities and films stripped to the waist, working vigorously alongside the workers. The victory over the Pontine Marshes was thus presented as a triumph of fascist ideology and engineering.[44]

Even today, the transformed region is one of the most prosperous in Italy. But the process was not without its casualties. First off, the relatively few, typically malaria-ridden hunters, fishermen and others who made their living among the marshes were driven out. Designated as 'derelicts of the outdated life on the land' by the project's chief engineer Ugo Todaro, these people were simply in the way, and in many ways the whole point of the project was to purge Italy of such undesirable elements, be they human or natural.[45]

The project employed tens of thousands of workers, peaking at 124,000 in 1933. Of these, untold numbers died of various causes. Though no official documents exist keeping track of the project's human cost, experts estimate that malaria alone took more than 3,000 lives.[46] Mussolini, who saw the project as a military endeavour, thought of lost workers as casualties of war, to be gladly sacrificed for the greater good. One of Italian fascism's most symbolic and conspicuous modernizing projects, the reclamation of the Pontine Marshes is a model of ruthless efficiency in bending nature to the will of man.

Owing to the lack of any environmental impact studies of the Pontine project, its ecological consequences are unclear, and are arguably offset by the progress it represented in fighting malaria in Italy. More recent swamp-draining projects, though, have had a more obvious and sinister impact.

Beginning in the 1970s, when Saddam Hussein first rose to power, the marshes of Iraq came under siege by his administration. Already suffering from private development and clearance, the marshes, home to the Ma'dan for thousands of years, became a focus of Hussein's early in his reign, as he ordered parts of them to be drained to accommodate military installations. During the first Gulf War in the early 1980s, he stepped up his campaign of draining the marshes for tactical reasons. Finally, in 1991, Hussein, angered at the Shia uprising with which many of the Ma'dan had sided, ordered a full-scale draining of the swamps as a reprisal against those who had struck against him and retreated to safety in the marshes. Through an aggressive programme of constructing drainage canals, locks and dykes, and even poisoning the waters to kill fish and animals important to the Shia, Saddam destroyed around 90 per cent of the Iraqi Marshes. The economic and environmental impact of the campaign is staggering.[47] Since the fall of Hussein's regime, a multinational coalition has made some progress in restoring some of the destroyed wetlands, but the culture of the Ma'dan has been largely destroyed, probably forever.

In Indonesia, we find one final, vivid example of the awful consequences of conquering the swamps. In central Kalimantan, the continuing, self-perpetuating fallout from a disastrous effort to tame and farm the swamps is striking and horrific. Here the swamps – or what remains of them – burn beneath the earth itself.

In the mid-1990s, Indonesia's President Suharto, acting against the counsel of environmental scientists, launched the Mega Rice Project, which would convert over a million hectares of peat swamp and forest into rice plantations. Workers moved in and dug a tremendous network of drainage canals, which opened up the formerly inaccessible swamps for logging. Agricultural efforts, both on the part of large plantation companies and small individual farmers, include extensive burning

in Indonesia, and the fires quickly raged out of control. Dried peat is pure fuel, and fires now smoulder in many places as deep as 60 cm (24 in.) below the surface, reigniting surface fires and sending massive amounts of carbon emissions into the atmosphere. Destroying these peat swamp forests can release drastically more CO_2 into the environment than other kinds of deforestation do – up to ten times as much, by some estimates. As a result, Indonesia is now the world's third biggest greenhouse gas emitter after the United States and China, due in large part to carbon emissions from peat swamp drainage and burning.

The nation has begun efforts to cut carbon emissions and to counteract the ecological toll of the Mega Rice Project and its fires, but the process, many fear, is self-perpetuating: prolonged, hot dry seasons due to climate change provoke and stoke fires, which contribute to further greenhouse gas emissions. This vision of swamps transformed into smouldering, subterranean fire fields is a sobering vision of the cost of 'conquering' swampland.[48]

Dead swamp in Louisiana with scorched trees in cotton valley field, killed by salt water and oil fire, 1926.

Of course, the true contemporary quagmire, the real challenge of twenty-first-century swamps, is how to protect and preserve the wetlands that nature and culture diminish with each passing year. Even where wetlands are protected and preserved by law, dangers ranging from drainage and development, to poachers hunting endangered species, to pollution, to natural phenomena like storms and wildfires – ever more menacing in an era of rapid climate change – threaten them. The wetlands of the Gulf Coast of the United States in recent years are a particularly poignant example – battered by hurricanes, polluted and poisoned to unknown degrees by the BP oil spill, itself only a particularly vivid and dramatic instance of a decades-long ongoing process of pollution, the coastal marshes of Louisiana and Florida are disappearing at an alarming rate. According to the United States Geological Survey, the last two hundred years

Cut-over Cypress Swamp, St Martin Parish, Louisiana, September 1985.

have seen the loss of around half of u.s. wetland habitats, owing to a combination of natural and human causes, and the loss of Louisiana wetlands, which account for 40 per cent of America's total wetlands, continues at a rate of around 75 km per year.[49] Essential to ecological balance as well as to their surrounding economies and cultures, wetlands almost everywhere are now much more fragile, shrinking resources than indomitable obstacles to human progress. The true problem they pose lies in determining how to protect and sustain them. As Walt Kelly's Pogo, cartoon denizen of the Okefenokee swamp, famously said, 'We have met the enemy, and he is us.'

3 Swamp as Horror: Monsters, Miasma and Menace

Swamps are apt breeding grounds for fear: dark, impenetrable and forbidding, they teem with life, much of it menacing. The treachery of swampy earth, a mix of land and water that threatens to swallow the unwary traveller at first misstep, compounds the threat. Alligators, serpents and uncertain depths aside, swamps foster terror in the way that deep, dark woods foster terror: like fairy-tale forests, they are the unknown, the uncharted, among the only spaces left where dark things might make their dens without discovery. But swamp horrors have a rich tradition of their own, beyond simple fear of the wild and unknown. Part of their special menace comes from their inherent mystery and uncertainty; genuine hazards, legend and imagination combine to transform swamp and bog, mire and morass into fitting backdrops for all manner of nightmares. Through history and across cultures, the swamp has been locus for an array of fears: some have been fanciful, the stuff of children's stories and B-movies, while others have been rooted in cultural guilt, haunting reminders of societies' sins against themselves, each other and nature.

Undeniably, much of the swamp's power to inspire terror comes from its longstanding status as outside the domain of civilization. Wilderness spaces in Europe and subsequently in colonial America have traditionally been associated with the demonic. In pre-modern Europe, the physical divisions between human settlements and untamed wild land became, symbolically, divisions between the terrestrial and the spiritual, whether

An alligator,
traditional terror
of the swamps.

divine or diabolical.[1] Any number of foundation myths feature
heroes conquering beasts, distillations of pure wildness, in order
to lay the groundwork for the spread of culture and civilization.
In an intriguingly swamp-centred example, the *Bibliotheca*, or
Library, a compendium of Greek mythological stories dating
back to the first or second century BC, presents Hercules meeting perhaps his most menacing foe in the swamp of Lerna: the
Hydra, a dread nine-headed beast who ventures forth from its
murky home to prey on the people of the plain.

The Hydra's strengths and dangers echo traits traditionally
tied to the swamp itself: it is inexhaustible, sprouting new heads
for each one the hero severs, like the stubbornly unconquerable
swamp; its very blood is deadly poison, like the miasmic respir-
ations of the swamp itself. Hercules is vexed by another swamp
denizen as he battles the Hydra, as an enormous crab lurking in
the muck attacks his feet. Columella, the Roman agricultural
writer, drew on this myth, dubbing the constellation Cancer
'Lernaeus' to commemorate Hercules' battle in the Lernaean
swamp. While Hercules ultimately triumphs, burning the stalks
of the creature's severed heads and then cutting it open to use its

blood on his own arrows, he must enlist the aid of his companion Iolaus to win the battle, and King Eurystheus refuses to count the labour towards his total ten. Even the mighty Hercules could not overcome the Hydra of the Lernaean swamp alone.

Gustave Moreau, *Hercules and the Lernaean Hydra*, 1876, oil on canvas.

While mighty, semi-divine heroes such as Hercules were often able to traverse and conquer otherwise untamable wilds, as well as the monsters they sheltered, the swamps, according to

legends including *Beowulf*, retained the ability to thwart even the judgement of God himself. *Beowulf*'s Grendel is one of literature's earliest marsh-monsters:

> Grendel this monster grim was called,
> march-riever mighty, in moorland living,
> in fen and fastness; fief of the giants
> the hapless wight a while had kept
> since the Creator his exile doomed.
> On kin of Cain was the killing avenged
> by sovran God for slaughtered Abel.
> Ill fared his feud, and far was he driven,
> for the slaughter's sake, from sight of men.
> Of Cain awoke all that woful breed,
> Etins and elves and evil-spirits,
> as well as the giants that warred with God
> weary while: but their wage was paid them![2]

Escaping the final judgement of God, Grendel and his mother become 'march-stalkers mighty the moorland haunting, / wandering spirits . . . Untrod is their home; / by wolf-cliffs haunt they and windy headlands, / fenways fearful'.[3] Grendel and the other creatures known as 'Cain's kin' were able to flee to and survive in swampy areas, waiting out the Flood thanks to their own 'semi-aquatic' natures.[4] As such, these creatures mark fens and marshes as spaces outside of God's domain, violating divine order, defying the judgement of the Flood and embodying all the evils and moral sickness that the Flood was meant to destroy. This conception of the wetlands makes them a fascinating reflection of the apparent limits of divine power – creatures of the swamp, who inhabit neither wholly land nor wholly sea, are able by their very in-betweenness to survive, to lurk, to fester, exiled in the swamps.

There is a longstanding literary tradition of turning to the swamps for moral metaphors. Dante's *Inferno*, for example, presents the Stygian swamp as a two-layered pit of damnation: those who were brought down by the sin of anger battle constantly in its mud and slime, tearing at each other with nails and teeth, while those beneath the waters suffer for a different sin:

Beneath the slimy top are sighing souls
Who make these waters bubble at the surface;
... Bogged in this slime they say, 'Sullen we were
in the sweet air made happy by the sun,
as smoke of sloth was smoldering in our hearts;
now we lie sluggish here in this black muck!'
This is the hymn they gurgle in their throats
but cannot sing in words that truly sound.[5]

'Grendel was said to have survived the Biblical flood by hiding in the swamps', J. R. Skelton's illustration from Henrietta Elizabeth Marshall's *Stories of Beowulf* (1908).

While the wrathful fight in the swamp, the slothful, the sullen, the creatures of inaction, lie mired beneath its surface. Connecting swamp-dwellers with laziness and moral lassitude is a common tendency that crosses cultures and eras. If progress towards civilization could traditionally be marked in terms of distance from the swamp, then those who remained must lack the industry to leave its easy but degraded bounties behind.

John Bunyan, in his 1678 Christian allegory *Pilgrim's Progress*, presents a sin-laden swamp in the miserable Slough of Despond, often translated as 'Swamp of Despair':

> And he said unto me, This miry slough is such a place as cannot be mended; it is the descent whither the scum and filth that attends conviction for sin doth continually run, and therefore it is called the Slough of Despond; for still, as the sinner is awakened about his lost condition, there ariseth in his soul many fears, and doubts, and discouraging apprehensions, which all of them get together, and settle in this place. And this is the reason of the badness of this ground.[6]

'Warriors carrying Grendel's head', J. R. Skelton's illustration from Henrietta Elizabeth Marshall's *Stories of Beowulf* (1908).

In Bunyan's work, sin itself afflicts the land, giving moral failing physical expression in an unnavigable and irredeemable mire that swallows carts whole. The swamps, then, become a metaphor for man's moral shortcomings, the physical analogue for the morass of sin in which the mortal traveller risks becoming lost. The associations of swamp and sin in two such foundational and influential religious texts as Dante's *Divine Comedy* and Bunyan's *Pilgrim's Progress* no doubt reinforced the tendency to look into the swamps' depths and perceive lurking evils.

William Blake's 1827
illustration of stygian
swamp-dwellers for
Dante's *The Divine
Comedy*, pen and ink
and watercolour over
pencil.

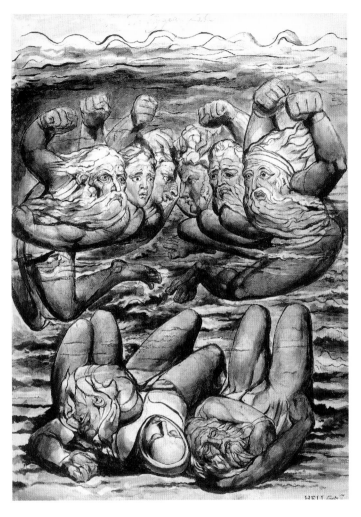

Wetland sites are also frequently associated with curses and supernatural punishment. Settlement sites that have sunk into boggy ground, or that have been overtaken by rising lake waters, often take on sinister significance in local legend. Archaeologists performing wetland digs frequently encounter sinister stories dating back centuries. Lore had it that an archaeological dig site at Lake Paladruin, France, might unearth a city that had been consigned to watery oblivion by a monk's curse. Another site, near Lake Valgjärr in Estonia, was said to be that of a manor house where a brother and sister had dared to marry in defiance of divine law; the manor was blasted by lightning, then the valley filled with rain and became a lake, consigning the entire marriage feast to oblivion.[7]

Fairy tales often ascribe a more explicitly diabolical menace to swamps, marshes and fens. In Hans Christian Andersen's fairy tales, swamps and marshes are home to the Devil and his kin, who wait beneath the surface to punish the prideful and unwary who venture into them. His story 'The Girl who Stepped on Bread' (sometimes translated as 'The Girl who Trod Upon a Loaf') centres on Inger, a proud, cruel, pretty girl given to torturing insects for her amusement. Her pride grows with her beauty, as does her cruelty. One day, as she walks home to visit her parents, the path she takes becomes increasingly muddy. Finally, as she comes to a great puddle, she tosses a loaf of bread into it and steps on it to avoid dirtying her good shoes. She and the bread sink together into the mud and slime of the bog, down, down into the lair of the bog witch:

> When the mist lies over the swamps and bogs, one says, 'Look, the bog witch is brewing!' It was into this very brewery that Inger sank, and that is not a place where it is pleasant to stay. A cesspool is a splendidly light and airy room in comparison to the bog witch's brewery. The smell that comes from every one of the vats is so horrible that a human being would faint if he got even a whiff of it. The vats stand so close together that there is hardly room to walk between them, and if you do find a little space to squeeze through, then it is all filled with toads and slimy snakes.[8]

As Inger sinks into the bog witch's lair, her body stiffens into paralysis, the bread still stuck to her shoe. As luck would have it, the bog witch is out at the moment, but the brewery is 'being inspected by the Devil's great-grandmother'.[9] The bog witch readily surrenders Inger to the Devil's great-grandmother, 'and that is the way she went to hell. Most people go straight down there, but if you are as talented as Inger, then you can get there via a detour.'[10]

While Inger's story ends with redemption, transformation into a bird and escape, the idea of the swamp or bog's porous, uncertain ground as a gateway to the underworld echoes in other fairy and folk tales as well. Another of Andersen's stories, 'The Bog King's Daughter', departs from the focus on a swampy demise as punishment for defects of character seen in 'The Girl who Stepped on Bread'. It tells of an Egyptian princess who uses a swan skin to fly to the great bog Vendsyssel, north of Jutland, to find a cure for her father's ailment. Andersen's description of the bog is intriguing, and evokes the sense that swamps and bogs are somehow primordial survivals of a wilder time:

> Today the bog is very large, but once it was even larger; there were miles and miles of swamp, marshland, and stretches of peat . . . over the bog hovered, almost always, a dense fog. At the turn of the eighteenth century there were still wolves there; and it was even wilder a thousand years ago.[11]

In a moment reminiscent of Persephone's capture by Hades, the Egyptian princess is taken by the bog king. She is not the first to be lost to the bog's depths: for centuries, when people walked on the bog's surface,

> they would slowly sink into the muddy ooze down to the bog king. That was the name given to the ruler of the great bog. Some called him the swamp king, but we prefer the bog king . . . Very little is known about his rule, which may be just as well.[12]

This mysterious figure steals his bride from the world above, and the rest of the tale concerns their offspring, a child who is a beautiful but cruel and nasty princess by day, and a sweet-natured but hideous frog by night. The dichotomy suggests, of course, that both spiritual and physical ugliness are the legacy of the bog king, while spiritual and physical beauty are the legacy of the selfless Egyptian princess who ventured into the swamp in search of a cure.

The theme of sinking through the swamp or bog to enter a supernatural netherworld reflects a common fear of swamp spaces – one grounded, to some extent, in reality. Wetlands all over the world show evidence of peoples having used them for burial grounds for thousands of years. Some burials seem to have been driven by a desire to preserve the dead, while others appear to have been motivated by ritual purgation: in Russia and China, suicides were often buried in bogs and swamps, for fear that their unquiet spirits would become ghosts. Records from eleventh-century Bavaria indicate that women who died in childbirth were sometimes buried in bogs or other unconsecrated ground for fear that the mother would return to take the child with her into her grave.[13] Bog bodies, or human remains preserved for centuries by the unique conditions of submersion in the peat, have been found throughout Europe, and their bodies hint at fascinating, often gruesome stories.

Bogs are uniquely suited for the preservation of bodies. Indeed, the discovery of bog bodies throughout Europe has been a boon to archaeologists and historians alike, offering incredibly well-preserved glimpses into the very distant past. For all the knowledge they can impart, though, the discovery of such bodies – or of parts of such bodies, dismembered in the process of cutting peat – can be a horrific experience. Bog bodies' preservation is uneven. While the flesh is often very well preserved, the bones typically become soft and spongy, and the internal organs, with the exception of the intestines, generally decay. Muscles, too, tend to deteriorate and shrivel beneath the well-preserved skin. These aspects of bog body preservation lend a particularly squishy, visceral horror to depictions of swamp and bog zombies in popular culture.

Bog bodies' preservation has sometimes led to confusion over dating, as the remains are often so well preserved that differences of centuries are imperceptible. In 1952 at Grabaulle in Jutland, Denmark, there was an ongoing argument between Professor P. V. Glob, an expert on bog bodies, and some of the

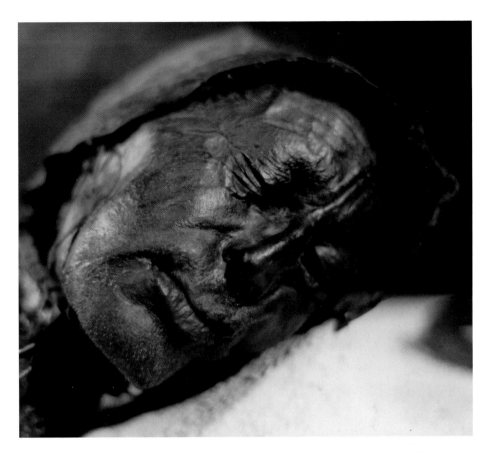

Head of Tollund Man, a mummified corpse found in a peat bog in Denmark in 1950. The body has been dated back to the 4th century BC.

locals about whether the bog body that had been discovered was centuries old, as Glob maintained, or the remains of 'Red Christian', a local peat cutter who had vanished in 1887, ostensibly after staggering into the bog drunk and drowning. The body was eventually found to have been dead more than 1,500 years.[14]

One notable event pertaining to the case of Lindow Woman, a famous bog body found in the Lindow Moss bog in Cheshire, England, is even more remarkable, and is in fact a chilling and ironic tale worthy of Edgar Allan Poe. In the early 1980s, two peat cutters came across a peculiar object in the bog. Cleaning it up, they discovered that it was a human head, remarkably well preserved, with some hair and one eyeball remaining intact. The

Southpoint forensics lab identified the owner of the head as a female between the ages of thirty and fifty.

When local man Peter Reyn-Barndt, who had formerly lived on the edge of the Lindow Moss bog, was told of the discovery, he was convinced that the jig was up. He quickly confessed to the murder of his wife, who had disappeared around two decades before. That confession was sufficient to convict him of murder – even after radiocarbon analysis showed that the skull dated back to the second century AD![15]

Don Brothwell, Professor of Archaeology at the University of York and expert on bog bodies, compares archaeological investigations of bog bodies to police forensic work, in that both lead frequently to evidence of terrible murders and brutality. Lindow Man, for example, shows evidence of an array of violent abuses: the skull was fractured in multiple places, evidently by a blunt weapon, and the remains of a cord tied around the neck and corresponding injuries indicate that he was strangled or garrotted. Evidence also indicates that his throat was slit, perhaps to drain his blood. Compared to other bog bodies, Lindow Man shows more, and more varied, injuries. The abundance of different kinds of injuries may suggest that Lindow Man was killed as part of an elaborate, violent ritual.[16]

Among bog bodies, the most common injuries found are skull fractures and evidence of asphyxia through hanging or strangulation. Tollund Man was found with a plaited rope noose around his neck, and Borre Fen man with a line around his neck. A female body found in Borre Fen had close-cropped hair, potentially echoing the common punishment for 'a guilty wife' among Germanic peoples of the time, which included chopping off her hair.[17] Evidence found in bog bodies indicates that punishment or ritual sacrifice persisted as late as Roman times. The Roman historian Tacitus reports that bogs were places of punishment among Northern tribes: 'Cowards, shirkers, and sodomites are pressed down under a wicker hurdle into the slimy mud of a bog.'[18]

Wetland sites have carried supernatural associations since prehistory. Archaeological evidence uncovered in wetlands around the world bears out the idea that ancient peoples regarded swamps and bogs as places where the human and spirit worlds connected. Sacrifices dating back to the Early Neolithic era have been uncovered in bogs around the world. Pairs of flint axes have been found in Sweden, buried together in a way that indicates that they were most likely votive deposits, or offerings to the divine.[19] Archaeological scholarship indicates that the bogs of Ireland were both prehistoric burial grounds and homes to a wide array of supernatural beings, whose attitudes towards humans ranged from neutral to malevolent. Trickster spirits like the pooka were said to bedevil those who lingered in the Irish bog; Huldre fen, in Jutland, is named for the Scandinavian equivalent, the *huldre*, a wicked spirit who tempted travellers into the bogs or marshes, never to be seen again.[20]

Legends and lore resonate with the visceral perception of swamps as lairs for wondrous beasts. Cultures all around the world tell of monsters or malevolent spirits dwelling in the depths of the swamps, both in ancient folk tales and in more contemporary legend. Swamp monsters typically fall into two major categories: the primordial survivors, the dinosaur-like leviathans who are supposedly preserved from a dark and savage past; and the humanoid lurkers, bestial, manlike creatures who range from seemingly peaceful missing links to wicked, even supernatural menaces.

Rising out of the swamps' primordial wildness, great beasts inhabit the legends of many countries. Typically reptilian, these creatures often seem to be natural exaggerations of genuinely imposing swamp fauna like alligators, snakes or giant turtles. Africa boasts several of these creatures, among the most famous being the Ninki-Nanka, a dragon-like beast said to inhabit the swamps of the Gambia and other parts of West Africa. The Ninki-Nanka's characteristics vary from account to account, but most agree that it is a giant serpent with a head crowned with either a horn or a diamond.[21] The threat of this creature, sometimes described as up to 10 m long, is often used to frighten children away from the swamps; the sight of it, supposedly, is enough to ensure that the viewer will die shortly after.[22]

Other African creatures in this category include the Kongomato, a bat-winged reptile said to inhabit the Bangweulu and Jiundu swamps of Zambia and Zimbabwe. There have been several supposed sightings over the years, and while most believe the sightings were merely birds or flying squirrels, the original inspiration for the Kongomato may have an intriguing explanation: one palaeontologist believed that 'the Kongomato tradition originated with natives who assisted in the excavation of pterosaur bones at the Tendagaru fossil beds in Tanzania prior to World War I.'[23] Belief in the Kongomato was strong enough that, in 1923, some of the region's Kaonde people 'carried amulets that would protect them from a Kongomato'.[24] Other dinosaur-like swamp creatures of African legend include the Mokele Mbembe, an elephant-sized beast said to have a horned snake's head at the end of a long, serpentine neck, purported to live in the swamps and rivers of the Congo, and the Chipekwe, a long-necked central African creature said to resemble a giant lizard.

These sorts of giant saurian are not limited to Africa. In the swamps around Ziro in India's Apatani Valley, a colourful

An American alligator, 13 ft (3 m) long, October 1988.

horned creature known as the Buru was said to prey on the local populace. An enormous lizard spanning 4 m from nose to tail, the monster was said to snare unwary people and pull them into its watery lair. Differing accounts and descriptions of the Buru exist. Ralph Izzard, who journeyed to the Himalayas in search of the Buru, recounts a description given to him by the inhabitants of a small Dafla village:

> During the rainy season, the swamp quickly floods to become a large lake; and when the water rises large animals appear from the swamp . . . They are much bigger than a man, and are bulky animals the same shape and size as the tame bison. They have small horns, but they point backwards instead of sideways like those of a bison. Like the bison they are colored black and white.[25]

As with all legendary beasts, consistency between accounts is rare. While both the Apatani and Nisi people described the Buru

A beast of the swamp.

to Western visitors in the 1940s, the Apatani believed the Buru was driven to extinction when they drained and cultivated the area's swamps.[26]

An American variant of the giant swamp-dwelling reptile is the Beast of 'Busco, a gigantic turtle said to live in Indiana's Black Oak Swamp. Named for the neighbouring town of Churubusco, Indiana, the Beast of 'Busco was first spotted by a farmer named Oscar Fulk in 1898, and has supposedly been sighted periodically since then by generations of inhabitants. Rumoured to be a giant alligator snapping turtle, the Beast of 'Busco has never been caught or photographed, and has escaped every attempt to capture it, including an attempt to drain the lake in which it lived. As with many American monsters, the Beast of 'Busco has become a local celebrity, feted each year at the local Turtle Days festival.[27]

The bestial humanoid is the second major category of swamp monster, though it encompasses a considerable array of types. One type is the witch or hag – a sinister, magical creature, depicted with varying levels of humanity or animalism. Unlike the reptilian leviathans, they tend to be unequivocally evil and prone to preying on children. One fascinating example is the Welsh crone known as the Gwrach-y-rhibyn, or the 'Hag of the Warning'. This creature was said to dwell in Wales's Caerphilly Swamp. When the river Nant-y-Gledyr was dammed, its waters pooled, becoming a practical defence, like a moat, for Caerphilly's castle. When the swamp waters pooled in this way, though, the Gwrach was said to rise from the inundated marshland.[28] The true terror of the Gwrach, beyond her hideous appearance, was what her coming meant for someone in the town, as the whisper of her wings always heralded death. Different accounts depict the hag in different ways; all agree that she had great wings like those of a bat, as well as talon-like claws on her fingers. The nineteenth-century folklorist Wirt Sikes provides an evocative description:

> The spectre is a hideous being with disheveled hair, long
> black teeth, long, lank, withered arms, leathern wings, and

a cadaverous appearance. In the stillness of night it comes and flaps its wings against the window, uttering at the same time a blood-curdling howl, and calling by name the person who is to die, in a lengthened dying tone, as thus: 'Da-a-a-a-vy!' "De-i-i-o-o-o-ba-a-a-ch!"[29]

While the Gwrach's manifestation primarily indicated an inevitable death, some tales also employ her as a cautionary figure. The Gwrach could alter her appearance, and even her gender, though by all accounts she is fundamentally a female who can choose to appear as a male (she would make an interesting study for more contemporary gender theorists). Sikes recounts a story of a drunken sot who encountered the Gwrach on the road in the darkness, and, mistaking the creature for human, 'made wicked and improper advances to it, with the result of having his soul nearly frightened out of his body in the horror of discovering his mistake. As he emphatically exclaimed, "Och, Dduw! It was the Gwrach y Rhibyn, and not a woman at all."'[30]

As is the case with many such beasts of local legend in the Western world, the Gwrach-y-rhibyn now functions more as

Alligator snapping turtle (*Macrochelys temminckii*) in the Oasis Park, La Lajita, Pájara, Fuerteventura, Canary Islands, Spain.

a draw for tourists than as a genuine fright. A website encouraging Caerphilly tourism plays on visitors' fascination with the supernatural and macabre in a colourful section devoted to the Gwrach:

> During every downpour when skies started to slobber, the hag would flap her bat-like wings and wail horrible enough to give everyone the habdabs. The hyena-hearted Gwrach would hover around the hoary mansions beating her wings . . . shrieking the names of those about to die.[31]

In the twenty-first century, the Gwrach-y-rhibyn, like most legends, seems to have lost her power to evoke genuine fear, though the death she predicted is no less inevitable.

The United States South played host to many swamp creatures and legends. They are, perhaps, unique among the world's swamp folklore in that so many of them grew out of the realities of slavery – some imported from African folklore traditions, others stemming from the trauma of slaves' experiences in

The Gwrach-y-rhibyn, a creature of Welsh folklore, from Thomas Croker, *Fairy Legends and Traditions* (1828).

America. Swamps were seemingly everywhere in the antebellum South, and they presented any number of challenges to white plantation owners whose livelihood depended on slave labour and the cultivation of land, as well as both hardships and opportunities for enslaved African Americans. For the slaves, the swamp had ambivalent meanings: it was both a menacing, forbidding place, fraught with the same terrors it conventionally held for everyone else, and a potential refuge, a place where an escaped slave might elude pursuers and live in comparative freedom. For all their terrors, the swamps provided enslaved Africans with a protected space, removed from the reach of institutional power, and with that limited freedom came subversive possibilities undreamt of on the plantation.[32]

The spectre of the swamp as a means of escape for slaves gave rise to a variety of kinds of swamp stories. Folk tales, passed around among the slaves themselves, described supernatural evils that made their homes in the swamp. The most notable of these is Jack O'Lantern, a frightening creature who lay in wait for unwary travellers on rainy nights and dragged them back to his home in the swamps. In an interview with the Federal Writers' Project in 1938, an ex-slave named Camilla Jackson describes the legend:

> During slavery and since that time, if you should go out doors on a drizzling night for any thing, before you could get back Jack O'lantern would grab you and carry you to the swamps. If you hollowed and some one bring a torch to the door Jack O'lantern would turn you aloose.[33]

Wirt Sikes describes a version of the same creature from African American folklore:

> They call it Jack-muh-lantern, and describe it as a hideous creature five feet in height, with goggle-eyes and huge mouth, its body covered with long hair, and which goes leaping and bounding through the air like a gigantic grasshopper. This frightful apparition is stronger than any man, and swifter

than any horse, and compels its victims to follow it into the swamp, where it leaves them to die.[34]

Jack O' Lantern and his ilk serve the traditional function of the monster-in-the-swamp story: they designate the swamps as dangerous places, the realm of supernatural evils, and discourage slaves from wandering too far afield.

Those slaves who did escape into the swamps gave rise to tales of entirely different kinds of swamp 'monsters', who struck sensationalistic fear into the hearts of newspaper and magazine readers of the antebellum area, Northern and Southern alike. Depictions of savage, sinister swamp-dwelling escaped slaves, devolved into a kind of primal wildness by their distance from civilization, became common in the media of the time. These figures were at times rendered as Romantic noble savages; at others, they were figures of terror. 'Swamp maroons', as they came to be called, became a fixture of Southern swamp lore.

The most famous representation of the swamp maroon came in a piece by the writer and illustrator David Hunter Strother, commonly known as Porte Crayon, that was published in an issue of *Harper's* in 1856. The piece, entitled 'The Dismal Swamp', centres on the author meeting Osman, an escaped slave long removed from civilization. Osman, while in his middle years, is powerfully built, 'gigantic', and has 'purely African features . . . cast in a mould betokening, in the highest degree, strength and energy'. His face has a look 'of mingled fear and ferocity'.[35]

The significance of the swamp maroon depended, of course, on one's attitudes and region. To abolitionists, he was a vision of elemental nobility and a reminder of the tragic dehumanization brought about by slavery. To white Southerners, he was a monster, a bogeyman, a reminder of the limits of slaveholders' control and a lurking threat of retribution.

Depictions of swamps in plantation fiction often emphasized, to scary effect, their distance from the ideal of gentlemanly control that defined the aristocratic South. John Pendleton Kennedy, in his archetypal plantation novel *Swallow Barn*, includes a lengthy scene in which his planter-class protagonists venture into the sinister Goblin Swamp. Kennedy's language in describing the swamp reflects their fear:

> Here and there a lordly Cypress occurred to view, springing forth from the stagnant pool, and reposing in lurid shade. Half sunk in ooze, rotted the bole and bough of fallen trees, coated with pendent slime . . . This dreary region was neither silent nor inanimate; but its inhabitants corresponded to the genius of the place . . . far from us, in the depths of darkness, the screech-owl sat upon his perch, brooding over the slimy pool, and whooping out a dismal curfew, that fell upon the air like the cries of a tortured ghost.[36]

The use of the term 'curfew' is telling, here, reminding us that the young men are venturing beyond boundaries, violating unspoken rules, in walking the swamps at night. Only through the aid of a swamp-dwelling guide named Hafen Block are the men led back to safety; from the comfort of the veranda, Hafen then treats them and the readers to a kind of fable about a man named Mike Brown who meets the Devil in the depths of the swamp, and falls prey to his deceptions.

While Kennedy's plantation idyll gives the swamp and its menace their due, a more telling swamp tale from the plantation genre comes in a collection from after the Civil War. In Thomas Nelson Page's In *Ole Virginia, or 'Marse Chan' and Other Stories* (1887), one tale stands out from the wistful idylls of a writer best known as one of the pre-eminent Old South apologists. Writing during Reconstruction, Page became a kind of literary ambassador for the South, presenting a romanticized plantation world meant to spark nostalgia in Northern and Southern readers alike. His was a South of Cavalier gentlemen, noble ladies and happy, loyal slaves; very seldom in his work are the horrors of slavery even obliquely suggested. 'No Haid Pawn' is an intriguingly swampy exception.

A ghost story that brings together swamp terrors both natural and supernatural, 'No Haid Pawn' combines hints of the supernatural with the terror of the maroon in an unsettling tale of violence and vengeance quite at odds with Page's usually harmonious plantation world. A dilapidated estate at the centre of a vile, ghostly swamp, 'No Haid Pawn', is shunned by all. The brave, adventurous narrator, son of a local plantation owner, has never dared venture within. Local slaves call it 'de evil-speritest place in dis wull',[37] and even escaped slaves would rather be recaptured than risk fleeing into its depths. The broken-down house at the swamp's centre has accrued an evil aura owing to an array of rumoured horrors: a slave was decapitated in an accident during its construction; many more were stricken by malaria while building it and cast into the murky swamp 'by the scores' before they were even dead. Page weaves together two stories: one, the story of the cruel, murderous West Indian slaveholder who once owned the place (his insistence on identifying this evil place with racial and cultural outsiders attests to his overall goal of ennobling the plantation South); the other, the story of a bold, powerful, recently escaped slave, 'as fearless as he was brutal'.[38] The narrator ventures foolishly into the wicked swamp, and is compelled by a sudden violent storm to pass a night in the 'haunted' estate. He is awakened in the night by cursing in a Creole patois. Believing the ghost of the homicidal owner has come for him, the narrator finds himself face to face with a terrifying vision:

> Directly in front of me, clutching in his upraised hand
> a long, keen, glittering knife, on whose blade a ball of fire
> seemed to play, stood a gigantic figure in the very flame
> of the lightning, and stretched at his feet lay, ghastly and
> bloody, a black and headless trunk.[39]

The horrific scene is blasted into oblivion by sudden lightning that destroys the house and allows the narrator to escape.

As interpreters of the story like Louis Rubin have suggested, the narrator's encounter is most likely with the escaped slave, who is known to have been slaughtering livestock for his own survival – hence the headless trunk. This explanation, while avoiding the supernatural, actually links the story more powerfully to the repressed guilt about slavery that Page's work generally ignores: the escaped slave, driven, like the others, by 'the cruelty of [his] overseers, or by a desire for a vain counterfeit of freedom',[40] prefers this blighted, pestilential place to his life on the plantation.

Most swamp humanoids tend to be less overtly menacing than the supernatural Gwrach-y-rhibyn, and less grounded in cultural guilt and injustice than the maroons of Southern legend. The United States boasts a number of these legendary creatures, who tend to be represented either as Bigfoot-like giants or as some form of lizard-man. Sightings of great, shaggy humanoid creatures have been reported in swamps in Louisiana, Michigan, Wisconsin, North Carolina and New Jersey, among others.[41] One of the most famous of these creatures is Louisiana's Honey Island Swamp Monster. This beast is said to stand 2 m (6 ft) tall and is most often described, in an intriguing postmodern variation on the trope of the primal missing link, as a 'Wookie', a name which merges folklore with Hollywood flash and marketability. Like the Beast of 'Busco and the Gwrach, the Honey Island Swamp Monster has become a tourist attraction, as an array of swamp tours of the area hint in their promotional materials at the possibility of a sighting.

As the Cajun Encounters Tours website explains,

> Legend has it that a large, ape-like creature with gray hair, red eyes, and a tepid disposition resides here in the Honey Island Swamp . . . Reports at the Cajun Encounters Camp vary from person to person, but those who have seen it say that [*sic*] darts about, more scared of us than we are of it.[42]

The first recorded sighting was in 1963 by a man named Harlan Ford, whose granddaughter now curates a website devoted to the beast. On the site are photographs of castings of tracks taken from the swamp, artists' depictions of the beast based on Ford's descriptions, and clips and pictures from interviews and documentaries about the monster. Most intriguing is a quote from Ford, describing his reaction to coming face to face with the creature:

I wouldn't have shot it unless I had to . . . Because it looked too human. But it wasn't human. It stood about 7 feet [over 2 m] tall, and had long, greyish hair hanging around it's [*sic*] face. I will never forget its eyes. They were large, amber colored and menacing.[43]

This category of beast, the swamp ape or wild man of the swamp, tends to be described in terms that emphasize its humanity as well as its otherness. As a beast man who is part monster and part human, he tends to be depicted as benevolent, or, at least, as desiring only to be left alone: a reminder of a lost wildness in our shared evolutionary past.

The unavoidable fact is that, in an era when the swamps themselves are far more threatened than threats, swamp monsters have lost much of their ability to frighten us. They have become curiosities of cryptozoology, mascots to drive tourism, harmless creatures invoked with a wink and a nudge by cultures beyond superstition. As we have come, owing to the march of progress, to view swamps as endangered spaces in dire need of preservation rather than as impenetrable, unnavigable realms of the dark unknown, swamp monsters in fiction and popular culture have come to reflect those changing attitudes.

Since the dawning of ecological consciousness in the mid-twentieth century, the nature of monsters has undergone a profound and inevitable change. When nature becomes threatened rather than threat, when wildness is menaced by progress rather than vice versa, creatures who embody or represent that wildness take on a very different character.

Most of popular culture's contemporary swamp monsters are protagonists rather than antagonists – embodiments of nature who serve to caution humankind about its abuses. Perhaps chief among the purveyors of contemporary mythology are comics companies. Each of the industry leaders, Marvel and DC, features a swamp-centric monster who, however horrific he may be in appearance, functions as a hero. Marvel Comics' Man-Thing, who debuted in 1971, is a biochemist named Ted Sallis who, on the run from nefarious villains, injects himself

Honey Island Swamp, Louisiana, rumoured home of the legendary monster.

with a top-secret formula and then crashes his car in the Florida swamps. Because he crashes close to the mystical Nexus of All Realities, a portal to alternate dimensions, Sallis's body is reconstituted from the vegetation of the swamp into a huge, mindless, shambling monster. Though mindless, for the most part, the Man-Thing feels the emotions of others – most notably fear, which causes him pain. Anyone feeling fear burns at the Man-Thing's touch. Frightening but not evil, the Man-Thing dwells within the Florida swamps, protecting his home from interlopers and threats and, in so doing, protecting the Nexus of All Reality. While Man-Thing's popularity has never rivalled that of Marvel's more conventional superheroes, he has developed a cult following – writer Steve Gerber's run on the title in the 1970s stands as a fan favourite, and contemporary writers often cite it as an influence.

DC Comics' Swamp Thing, while similar to Man-Thing, has enjoyed greater popularity and critical acclaim. The two characters debuted in the same year, 1971, and their origins, while wildly offbeat, are remarkably similar. Alec Holland, a scientist working in the Louisiana swamp, died in the toxic, polluted waters and was reconstituted from plant matter – his 'spirit was enveloped in a plant body that thought it was still human'.[44] The newly constituted 'plant elemental' is sentient, and, unlike Man-Thing, quite intelligent; in addition, it is superhumanly strong, can regenerate itself and has a variety of other powers ranging from supernatural awareness to time travel. The writer Alan Moore, the creator of such series as *Watchmen* and *V for Vendetta*, who has long been synonymous with the maturing of the comics

Dick Durock as Swamp Thing in the 1982 film.

genre, transformed the character in the 1980s on a critically acclaimed run that brought out both an elaborate mythology involving earth spirits and avatars and an even stronger environmental theme, as he presented the sentient plant-being as a 'guardian of the Green', defending the balance of nature against evil and exploitation, or 'The Red'.

These contemporary swamp monsters are, like those before them, reflections of the culture that produces them. The ecological messages driving Man-Thing and Swamp Thing have been more or less overt over the creatures' forty-plus years in the comics, and their fundamental identities as heroic monsters, embodying pure nature, battling against those who would pollute, defile or pervert it, signal fundamental changes in attitudes towards swamps as endangered wetlands, emblematic of a distressed and embattled nature as a whole.

Horror films, too, provide an intriguing variety of perspectives on the swamp and the fears it evokes. A wealth of horror movies of varying levels of quality and taste turn to the swamps for their setting and inspiration. Swamp horrors in film range from the formulaic to the strikingly unconventional, and exploit the timelessness and ambiguity of the swamp setting to play on fears ranging from the primal to the post-apocalyptic.

Many swamp horror films are of the B-movie variety – low-budget, lowbrow affairs whose ubiquity may make them more telling measures of mass cultural trends than critical acclaim or artistic merit might. Movie studios have churned out innumerable swamp-based horror films since the mid-twentieth century. These films tend to emphasize one of, or a combination of, a few themes. Most common, of course, is the swamp monster – the giant alligator, the vaguely reptilian humanoid or the otherworldly monstrous presence who preys on the unwary visitor. This theme's origins are fairly obvious, as it simply exaggerates the snakes and alligators one might actually find if one ventured into the swamps, and plays on the swamps' sense of primordial menace. Another common theme in swamp horror, perhaps inspired by both the swamps' ghostly associations and the legacy of curses and secret sins so prevalent in swamp legends, is the return of the vengeful killer. Typically, the villain will be someone wronged, who returns for justified, if misguided, vengeance. In *Strangler of the Swamp* (1946), the killer is the spirit of a wrongly condemned man. In *Bikini Swamp Girl Massacre* (2014), the killer is the spirit of a Seminole warrior seeking vengeance for his people. Perhaps not surprisingly, the film has little of substance to say about the plight of the Native American.

Blending scientific hubris with primordial peril, the archetypal mad scientist pops up frequently in swamp horror films. Often rejected by the scientific community or even by civil society as a whole, the mad doctor, either misguided or outright malevolent, uses the swamp as a site for his unspeakable experiments. Manuel Caño's *Swamp of the Ravens* (1974) deals with this theme, as a mad doctor attempts to reanimate corpses in the midst of a swamp filled with monstrous bird-beasts. In the emphatically titled *Swamp Zombies!!!* (2005), a doctor, experimenting with a method to resurrect the dead, ditches his failed experiments in a nearby swamp, with predictable results. The juxtaposition of misguided science with swamp wilderness underscores, in films like these, the folly of going too far in defying the laws of nature, as the results of the experiments – undead hordes, misshapen humanoids, monstrous reptiles – inevitably turn on their creators.

Of these mad-scientist-driven films, the camp classic *Curse of the Swamp Creature* (1966) is a particularly interesting example, though a film deeply suspect in both aesthetic quality and taste. It deals with a scientist whose experiments centre on evolution; he has gone to the swamps because of the abundance of snakes and alligators there, and has managed to create an enormous and frankly rather comical humanoid monster. What makes the film both interesting and distasteful, however, is the representation of the African American 'natives' who live among the swamps. They warn him of coming danger through a system of jungle drums, and they are depicted doing strange tribal dances in masks around a totem, as the film blends jungle and swamp clichés to make the most of the alienness of its setting. The film also plays on viewers' fears of rural folk, of being cut off from civilization and left to the mercies not just of a monster and a mad doctor, but of strange, dangerous country people who may as well have come from the jungles of Africa as from the u.s. South. Most swamp horrors of this ilk tend to take an array of scientific, historical and cultural liberties in their depictions. The very alienness of the swamp landscape for most viewers, combined with the generally low expectations of the genre, gives the film-makers broad licence; thus it is not unusual to see a wildly misplaced animal species or gross cultural inaccuracy pop up in them, as in the case of *Curse of the Swamp Creature*'s tribal natives.

Another frequent source of horror in swamp films is, of course, the sense of separation from civilization and control that swamp settings can bring. There is an entire subgenre of horror films that deals with the fear of out-of-the-way places, and of country folk warped and distorted by their distance from mainstream culture. This genre is not limited to swamps, of course: the Wes Craven classic *The Hills Have Eyes* (1977) dealt with the horrifically transformed victims of nuclear tests in the desert; the *Wrong Turn* film series (2003–14) features hapless travellers stranded in the forests of West Virginia and hunted by bizarre cannibals. While not strictly a horror film, *Deliverance* (1972) has gained an indelible place in popular culture for its terrifying depictions of rapacious, murderous hinterlanders. While this genre raises some troubling questions about class and regional prejudice, the fear of being stranded among strange, dangerous people in an inhospitable wilderness is elemental.

While many such swamp horror films are set in the American South, a recent notable example of the formula playing out in a different location is the 2010 film *The Reeds*, which is set and filmed in the Norfolk Broads area in England. The Norfolk Broads is an area of connected rivers and lakes, rarely more than a metre deep, that are generally navigable. A popular boating destination, the Broads, like other swampy areas, are both beautiful and potentially threatening – vast, full of high reeds that in some places seem impenetrable, and disorienting. *The Reeds* touches on several swamp horror staples. A group of young people on a holiday trip to get away from it all encounter strange, menacing locals in the form of teenage vandals and the sinister owner of the boat shop. As their trip progresses, their boat gets caught on an obstruction, and they begin to see horrific and ghostly images, including premonitions of their own deaths. Ultimately, the group falls prey to the curse of the moors, damned to wander the waterways for all eternity because of a past series of murders. Filled with disorienting close-up shots of the obscuring reeds as well as overhead views emphasizing the vast emptiness of the Broads, *The Reeds* is another horror film made possible by its setting.

While many swamp horror films tend to return to a relatively small range of themes and formulas, some film-makers make more ambitious use of the swamp setting's potential. Kaneto Shindo's 1964 film *Onibaba* (Demon Witch) is set in a vast reed swamp in feudal Japan. The film is an intriguing mix of genres: it is as much psychological drama as it is horror story. Drawing on Japanese mythology, it touches on the archetype of the witch in the swamp and hints at the supernatural while remaining rooted in the intensely human themes of sexual desire and jealousy and the dehumanization of war. Throughout, the marsh setting is thematically and visually essential.

The film begins with a hole in the marsh – 'The HOLE. Its darkness has lasted since ancient times,' reads the text over the film's opening shot. Its opening moments linger on the reeds, endless waves of dagger-like grasses through which soldiers struggle on foot while riders pursue them. A spear comes through the obscuring wall of reeds, killing the soldiers without warning. Once the soldiers are dead, two women, one middle-aged, one young, emerge, strip the victims of their valuables and then drag them off into the reeds, ultimately dropping them into the Hole. This is our introduction to the film's protagonists, credited only as Kichi's mother and Kichi's wife, named for their relationship with a character who has been killed in the war, and who never appears in the film.

The world of the film is one of desperation. Feudal war has left the people starving as men are impressed into service, leaving no one to till the fields. The cities have been burned, and people kill for food and clothing. When Hachi, a friend of the middle-aged woman's slain son, returns, he wears the clothes of a murdered priest. People trust a priest, he explains.

Even in its fourteenth-century setting, *Onibaba* is oddly post-apocalyptic. Cities have burned; food is scarce. We see the murderous mother and daughter-in-law at the film's centre chasing down a dog to kill and roast it. Many references to purgatory and to lost souls, coupled with close-up shots of the endless thickets of bladelike reeds, sharp enough to draw blood, make us question whether or not we are already there.

The central story draws on Japanese mythology, adapting a folk tale about a jealous woman punished for masquerading as a demon to frighten her daughter-in-law. Hachi and Kichi's wife begin a sexual relationship, which drives Kichi's mother mad with envy and lust. She eventually encounters a strangely masked samurai in the marsh one night while the two are making love. The samurai tries to force her to lead him out of the swamp, but

Kichi's mother, trapped in the demon mask, in *Onibaba* (1964).

she tricks him into plunging to his death in the Hole. Stealing his mask, she discovers that his face beneath is scarred, but nevertheless she dons the mask herself. She plans to frighten Kichi's wife away from Hachi. She ends up pursuing the younger woman through the marsh, but realizes that she cannot remove the mask. Kichi's wife helps her to break the mask away with a hammer, only to reveal that Kichi's mother's face is horribly scarred underneath. Terrified of the woman she thinks has become a monster, Kichi's wife flees through the marsh. She leaps over the Hole; the movie ends as Kichi's mother, running blindly, jumps after her, her existential fate sealed whether or not she plunges into the abyss.

Throughout, the film uses the otherworldly, post-apocalyptic swamp landscape to build an atmosphere of thematic terror. We never move beyond the sea of reeds. The dislocation, the sense of disorientation created by the rustling grasses is compounded by the vague references to the ruined world outside, with its burned cities and war-scarred wastes. The few characters we meet are amoral at best, driven to murder and scavenging by the simple need to survive. In an interview, director Kaneto Shindo has said that the mask's disfiguring effects on the samurai and woman symbolize the horrific effects on post-war Japan of the atomic bombs dropped on Hiroshima and Nagasaki. The timelessness of the marsh setting, with the eternal Hole at its centre, enables the timelessness of the film's metaphors.

Turning to television, and the 2014 HBO series *True Detective*, we see a more recent example of a swamp setting as source of both cultural commentary and existential horror. Like *Onibaba*, *True Detective* works within several genres, the pulpy overtones of its title notwithstanding. The story of two policemen attempting to solve a series of bizarre murders over a period of decades, *True Detective* plays out against the backdrop of a blighted coastal Louisiana, where coastal wetlands are disappearing at an alarming rate, a process exacerbated by the oil industry's channels and pipelines. The show's first season combines its crime story with elements of existential horror, enabled by the ephemerality of the coastal swamp setting.

Much has been made of the literary influences on *True Detective*, particularly that of existential horror. The spectre of the Yellow King; the idea of dark, occult societies; secret, perhaps satanic churches; defilement of innocence; walking monsters; the hell-on-earth that is Carcosa – these all echo, if not outright embrace, the darkly supernatural. The Yellow King comes from a story not far removed from the Cthulhu mythos of H. P. Lovecraft, conjuring an image of slumbering, malevolent forces beyond comprehension that threaten to burst forth if awakened and unleash chaos and destruction. In *True Detective*, actual catastrophic storms Andrew, Katrina and Rita, the hurricanes that savaged the Gulf Coast in the 1990s, function this way, wreaking havoc on the landscape, causing senseless, incalculable damage, and wiping out the records and remnants of truth that could help see justice done; even the swirling symbol that marks the Yellow King's disciples evokes the pattern of a hurricane. The real horror of *True Detective* is equal parts noir and cosmic, equal parts Kafka and Cthulhu, as the good ole boy bureaucracies, petty venalities and self-serving defences of the status quo shield and shelter the unspeakably dark acts to which they turn a blind eye, while nature batters and inevitably swallows the compromised landscape itself.

Of course, it is not nature alone that threatens the landscape. In *True Detective*, the machine has despoiled the garden, and the show conveys a pervasive sense of impermanence and destruction: as Rust Cohle observes at one point, 'Pipeline covering up this coast like a jigsaw. Place is going to be under water within thirty years.' Director Cary Fukunaga presents an expressionistic landscape full of loss, tragedy and menace – conscious of the ecological toll exacted by industry, and resigned to it. The real horror of *True Detective*, then, has as much to do with ecological devastation as it does with murder.

Whether embodied in ancient myth, regional fiction or contemporary comics and films, swamp monsters emerge throughout history as indicators of cultural values, fears and anxieties – the swamp itself, as a kind of repository for the cultural unconscious. In all, swamp monsters, however dark and mysterious, embody visceral and foundational elements of human nature. They reinforce our borders, guarding the ever-shrinking realms of the forbidden and unexplored, waiting to be conquered by founding heroes or to punish those who overstep their boundaries. They hint at our origins in saurian survivals of a primordial world, or shaggy hominids wandering our few untamed spaces; they remind us of our mortality in black-winged crones whose wingbeats herald death; they embody our individual and collective sins as menacing maroons driven utterly bestial by a practice that already robbed them of their humanity, or as human-plant hybrids who guard nature against our abuses. While the Devil's grandmother or the swamp king may lurk beneath the murky surface, the real horrors of the swamp come from its reflections of us: of our history, of our failings, of our troubling future.

4 Swamp as Spectacle: Paradise, Sanctuary, Aesthetic Object

There's something about being out in the swamp, especially when you're alone, that seems to suck the poisons of civilization out of you ... some sort of healing power that ... lets you begin to be yourself again.

Photographer and author Greg Guirard, *Atchafalaya Autumn* (1995)

The acclaimed BBC television series *Planet Earth* captures an incredible and moving vision of the Okavango, the great African river delta. At the peak of the dry season, the enormous wetland system known as the Jewel of the Kalahari becomes suffused with waters from faraway rains that have flowed great distances to collect here and return a bleak, dry landscape to glorious life. In the Okavango Delta, fresh water leaks from river channels into back swamps, in part through hippopotamus-created passages beneath the papyrus connecting rivers with lagoons and other bodies of water, or is redistributed among inundated areas through surface channels created by elephants and other large animals. As these waters return, a vast and varied array of animals trek for weeks to find water in the Okavango Delta – zebras, giraffes, impala, antelope, baboons, all make a kind of annual pilgrimage to this source of sustenance, this ultimate oasis. In all, the delta sustains 'more than 400 species of birds, about 1300 plants, 71 fish and tens of thousands of invertebrate species'.[1] Framed as the film-makers of *Planet Earth* present it, the Okavango Delta becomes a vision of paradise – an unspoiled, life-giving Eden, imparting life to the magnificent creatures of the Kalahari.

The Okavango Delta contains a variety of swamps, ranging from a large area of permanent swamp to other areas that are transformed into swamps by seasonal or occasional flooding.[2] These swamps are essential to the delta's regenerative and nourishing function: 'It is mainly in the seasonal swamps and

View down into the
Okavango Delta.

occasional floodplains – where the aquatic and terrestrial worlds meet – that the Delta comes to life, and its wealth is produced.'[3] The waters in these swamps are clean and unpolluted, so much so that, as the film claims, 'anyone visiting the Delta or using its water must be astonished at its clarity and purity.'[4]

Sunset over the Okavango River.

South America's Pantanal, a vast, varied wetland system that spans parts of Bolivia, Brazil and Paraguay, and whose name essentially means swamp or marsh,[5] is another such jewel, similarly celebrated for its purity. The Pantanal is an extraordinary and unique ecosystem, alive with hundreds of kinds of birds, thousands of different butterfly species, a magnificent array of flora and even jaguars, the signature animal species with which the area is associated. Scholars writing of the area emphasize not only its incredible beauty and biodiversity but its status as a pristine, vulnerable Eden; one scholar goes so far as to call it the 'one paradise left' in the wake of the damage humans have inflicted upon nature.[6]

Such Edenic visions of these swamps – of any swamps – represent a sharp departure from traditional swamp representation.

As we have seen, through most of history swamps have been regarded negatively, whether owing to tradition and trope, to practical struggles with swamps as obstacles or hazards, or, most often, to some combination of these factors. Though they have traditionally been depicted as places of death, swamps teem with life, nurturing flora and fauna. Long considered hotbeds of pollution, pestilence and disease, swamps actually clean and filter the waters that pass through them. We now know enough to see past the images of swamps as regions ruled by death and decay; most contemporary representations of swamps, while they may still draw on picturesque gloom or spooky menace, emphasize the beauty and fragility of these too-often endangered spaces. Part of this shift in perception is owed, clearly, to the fact that swamps are no longer the insurmountable, indomitable spaces they once were: when a space has to be actively protected, it loses much of its power to inspire fear and dread. Beyond simply losing our fear of swamps, however, we have come to regard them in ways largely antithetical to traditional ideas that linked swamps with pestilence and evil.

Pantanal in Mato Grosso, Brazil.

At the heart of the swamps' transition in popular imagination from poisonous hell to unspoiled Eden is a series of philosophical and cultural shifts in attitudes towards nature and civilization that replace old ideas and associations with new. Swamps have largely come to be celebrated as places of pure wildness, of prelapsarian innocence: unspoiled (but always, and increasingly, threatened) antitheses of corrupt civilization, sprawl and blight. This alternative vision comes partially from scientific and technological progress: we understand the swamps more fully and can, and do, drain and clear them with alarming ease. It also owes a great deal to evolving attitudes about nature and wilderness in general. The rehabilitated, romanticized swamp that emerges blends with persistent, prevailing swamp stigmas to add new dimensions to the swamp's multivalent complexity.

Throughout much of Western history, nature itself has been suspect, if not outright sinister. The wild, the untamed,

Turtles, Lake Charles, Louisiana.

in landscape signified moral dissipation or waywardness in humanity, and swamps, as depicted in art and literature, nearly always signified chaos, danger or wickedness. Most colonists in America, of course, cast the wilderness as evil in no uncertain terms: while the English Romantics were influenced by the otherworldly, allegorical visions of Dante and Milton, American attitudes towards nature had been shaped by a Puritanical condemnation of the wilds rooted in practical fears and demonization of the savage. Many early American texts, in particular captivity narratives, use the swamp as a metaphor for the trials of the soul, calling up references to the underworld, echoing tropes of death and sorrow. Native Americans were often depicted as swamp-dwelling devils, as in the Puritan William Bradford's description of a meeting of medicine men, who gathered 'for three days together in a horrid and devilish manner, to curse and execrate them with their conjurations, which assembly and service they held in a dark and dismal swamp'.[7]

This grim attitude towards nature, and towards swamps in general, persisted well into the nineteenth century, and still colours some swamp representations in popular film and fiction today (though rarely, in recent decades, without some countervailing ecological consciousness). Even Romantic visions of the swamps emphasized their melancholy and link with death. Perhaps the best-known swamp poem of the early nineteenth century was the Irish poet Thomas Moore's 'The Lake of the Dismal Swamp', a ballad which tells the tale of a young man whose love, buried in a grave 'too cold and damp / for a love so warm and true', goes to the titular lake to haunt the swamps in her white canoe. The young man, determined not to let her go, braves the swamp's hazards to reunite with her, and ultimately joins with his 'death-cold maid' in her white canoe. However Romantic the conceit, the poem presents the swamp as unambiguously deathly and menacing, making the youth's journey and sacrifice eminently more noble.

In North America, proximity to untamed swamps typically bred no more affection, particularly in the swamp-dense South. The South Carolina poet, novelist and historian William

Thomas Moore,
after a painting by
Thomas Lawrence,
19th century.

Gilmore Simms epitomizes the negativity in his poem 'The Edge of the Swamp' (1853):

> 'Tis a wild spot, and hath a gloomy look;
> The bird sings never merrily in the trees,
> And the young leaves seem blighted. A rank growth
> Spreads poisonously round, with power to taint
> With blistering dews the thoughtless hand that dares
> To penetrate the covert. (11.1–6)

Simms's swamp is melancholy, nigh-impenetrable, poisonous, potentially infectious. Its inhabitants are horrific – 'The cayman – a fit dweller in such home' is a 'steel-jawed monster' who 'crawls slowly to his slimy, green abode'. Others, less immediately threatening, are linked with death, or undeath: 'A whooping crane erects his skeleton form, / And shrieks in flight' (II.II–I2). Simms's terrors are not limited to the swamp's animal inhabitants. The vegetation itself exudes both mournfulness and menace:

'A Peep into the Great Dismal Swamp', illustration from Edward King's *The Great South* (1875).

Nothing of genial growth may there be seen,
Nothing of beautiful! Wild, ragged trees,
That look like felon specters, – fetid shrubs,
That taint the gloomy atmosphere, – dusk shades,

That gather, half a cloud, and half a fiend
In aspect, lurking on the swamp's wild edge. (II.33–8)

Simms's poem is a veritable catalogue of classic swamp images:
strange menacing beasts, hints of supernatural evil, a sickening,
poisonous atmosphere, even hints of criminality ('felon specters')
and the outright diabolic ('half a fiend / In aspect'). To amplify
the environment's menace, he describes the doomed journey
of a butterfly, defined by beauty and innocence, a 'dandy of the
summer flowers and woods' which to this point 'has counted
climes / Only by flowers', and which has drifted unawares into
this hellish morass. It alights on the alligator's head; the beast

October sunrise
in the Cocodrie
Swamp, 1987.

'Cypress swamp on the Opelousas Railroad', illustration from *Harper's Weekly* (8 December 1866).

submerges, making the butterfly '[dip] his light wings, and [spoil] his golden coat, / with the rank water of that turbid pond' (ll.28–9). The poem's closing image – the despoiled butterfly, now 'sad', fluttering away to more hospitable surroundings – serves as warning for the speaker and his party 'to speed / for better lodgings, and a scene more sweet, / Than these drear borders offer us to-night' (ll.42–4).

No hint of the beautiful, no hint of appreciation of or respect for the wild emerges in 'The Edge of the Swamp'. The tension in the poem comes from the contrast between the butterfly, exemplar of beautiful, fragile nature, and the menacing, miasmic swamp. Here, the butterfly is alien; even as it escapes, its beauty has been spoiled by the swamp's toxic waters. Any fascination the swamp holds lies in its terror; truly dismal, it must be shunned by wary travellers, duly chastened by the fate of the poor butterfly. Significant, too, is Simms's emphasis on the swamp's wildness – three times over the course of this short poem he reminds us that the swamp is wild, and that wildness, he implies, is the source of its menace. Simms's poem distils nearly all the conventional negative swamp associations against which more redemptive depictions would struggle.

Though the intellectual and artistic movement known as Romanticism, which took hold in different ways throughout Europe and the United States in the early nineteenth century, sought to replace Enlightenment-era rationalism with a new celebration of nature and wildness, swamp representations in English Romantic poetry remained fairly conventional, playing off traditional themes of gloom, sin and sorrow. Prominent English Romantic poets like John Keats, Percy Bysshe Shelley and Samuel Taylor Coleridge still tended to use landscape 'to signify moral and spiritual waywardness', or 'as a correlative of psychic disintegration'.[8] In America, though, swamps that were essentially symbolic and allegorical in Europe were physical facts, immediate practical challenges, sensory experiences to be encountered; thus American Romantics were at the forefront of a new vision of nature that came to celebrate the long-demonized swamps.[9]

The revolutionary impulses behind American Romanticism prompted a self-conscious rejection of inherited wisdom, perhaps especially as it regarded religion and conceptions of nature. Ralph Waldo Emerson, perhaps the most influential American Romantic, called both for a new American scholar who relied on his own, unmediated experience with the world around him rather than on the lifeless words of inherited, codified learning, and for a new relationship with divinity attained through contemplation of the natural world. In his famous essay 'Nature' (1836), Emerson asks,

> Embosomed for a season in nature, whose floods of life stream around and through us, and invite us by the powers they supply, to action proportioned to nature, why should we grope among the dry bones of the past, or put the living generation into masquerade out of its faded wardrobe?[10]

Truth, enlightenment, transcendence, even God, are to be found in nature itself:

In the woods, we return to reason and faith. There I feel that nothing can befall me in life, – no disgrace, no calamity, (leaving me my eyes,) which nature cannot repair. Standing on the bare ground, – my head bathed by the blithe air, and uplifted into infinite space, – all mean egotism vanishes. I become a transparent eye-ball; I am nothing; I see all; the currents of the Universal Being circulate through me; I am part or particle of God . . . I am the lover of uncontained and immortal beauty. In the wilderness, I find something more dear and connate than in streets or villages. In the tranquil landscape, and especially in the distant line of the horizon, man beholds somewhat as beautiful as his own nature.[11]

Emerson's ideas would have a profound influence on conceptions of nature and divinity in America, as would those of a writer and thinker whom he mentored, Henry David Thoreau.

Denizens of the Ocklawaha river, *c.* 1926.

Stephen Schoff after
Samuel Worcester
Rowse, 'Ralph Waldo
Emerson', 1878,
engraving.

'I wish to speak a word for Nature, for absolute freedom and wildness, as contrasted with a freedom and culture merely civil – to regard man as an inhabitant, or a part and parcel of Nature, rather than a member of society.'[12] So begins Henry David Thoreau's essay 'Walking' (1862), which is as succinct a manifesto of his attitudes towards nature as any work in his oeuvre. Thoreau is best known for *Walden* (1854), his account of his retreat to the nature and solitude of Walden Pond in an effort to 'live deliberately . . . to live deep and suck out all the marrow of life'. Such a spirit enlivens 'Walking' throughout, and gives context to his discussion of the swamps.

For Thoreau, wilderness is not an element in, but an escape from, the artificial constructions of human culture. His treatment of the swamp, specifically, exemplifies the contrast between experience shaped by convention and unmediated immersion in the wild. The swamp first appears in his essay in a negative light, as part of Thoreau's perception of a man unable to see nature except as possession and commodity: he imagines

> some worldly miser with a surveyor looking after his bounds, while heaven had taken place around him ... he did not see the angels going to and fro, but was looking for an old post-hole in the midst of a paradise. I looked again, and saw him standing in the middle of a boggy, stygian fen, surrounded by devils, and he had found his bounds without a doubt, three little stones, where a stake had been driven, and looking nearer, I saw that the Prince of Darkness was his surveyor.[13]

The stygian fen here is not a product of nature, but of the cheapening of nature through fences and boundaries – a hell of the miser's own creation.

Later in the piece, the swamp's pure wildness represents a stark contrast to such a condemnatory vision:

Martin Johnson Heade, *The Marshes at Rhode Island*, 1866, oil on canvas.

Hope and the future for me are not in lawns and cultivated fields, not in towns and cities, but in the impervious and quaking swamps . . . I derive more of my subsistence from the swamps which surround my native town than from the cultivated gardens in the village . . . Yes, though you may think me perverse, if it were proposed to me to dwell in the neighborhood of the most beautiful garden that ever human art contrived, or else of a dismal swamp, I should certainly decide for the swamp. How vain, then, have been all your labors, citizens, for me![14]

Autumn in an Everglades wilderness preserve.

The swamp, for Thoreau, feeds the soul in ways that garden cannot. Cultivation robs nature of its moral purity, of its ability to provide spiritual sustenance – a clear inversion of the Puritan ideal. Throughout 'Walking', Thoreau imparts a fundamental moral purity to nature, figuring nature as a nearly literal moral compass: 'I believe that there is a subtile magnetism in Nature, which, if we unconsciously yield to it, will direct us aright.'[15] The wilder the environs, the more distant from tame, docile civilization, the more genuine and pure the spiritual experience for Thoreau; and so, the swamp, wildest of landscapes, becomes for him a locus of spiritual renaissance:

> When I would recreate myself, I seek the darkest wood, the thickest and most interminable, and, to the citizen, most dismal swamp. I enter a swamp as a sacred place, – a *sanctum sanctorum*. There is the strength, the marrow of Nature. The wild-wood covers the virgin mould, – and the same soil is good for men and for trees . . . In such a soil grew Homer and Confucius and the rest, and out of such a wilderness comes the Reformer eating locusts and wild honey.[16]

Wild soil, swamp soil, gives birth both to intellectual and spiritual leaders, while cultivation despoils and corrupts. Thoreau's essay represents a radical shift towards undoing the idea of civilization as movement out of the swamps, an idea we have seen since the Scandinavian tribes moved from bog to hall, and heralds a new celebration of wilderness fundamental to a reimagining of swamps as nearer Eden than the underworld.

Radical as Thoreau was, he was understandably somewhat ahead of his time in embracing the swamps. Even among contemporary environmentalists, his assertion in a journal entry that it would be 'a luxury to stand up to one's chin in some retired swamp for a whole summer's day, scouting the sweet-fern and bilberry blows, and lulled by the minstrelsy of gnats and mosquitos' is unlikely to spark enthusiastic agreement. More common and widespread was a more measured re-evaluation of the swamps, one that valued them not so much for their purity as for their picturesqueness – an elevation of swamps not as Edenic spaces, but as sites of fascination and mystery, of danger and adventure. Writers and artists came increasingly to appreciate the natural beauty of the swamps, but that beauty gained a fascinating edge from the darker notions and superstitions surrounding it.

These mid-nineteenth-century changes in the ways that poets, travel writers and tourists regarded the swamps were reflected in the visual arts, as well. Landscape art had a longstanding tradition of casting nature as something to be transcended or tamed rather than 'communed with' – a notion that would have suggested all manner of moral depravity and sensuality. Romantic notions of nature began to supplant traditionally conservative, moralistic and nationalistic depictions of

A. R. Waud, 'Cypress Swamp on the Opelousas Road', *Louisiana Harper's Weekly*, 8 December 1866.

Soldier resting in a swamp, illustration from Sir James Edward Alexander's *Transatlantic Sketches* (1833).

nature as something to be conquered. Swamps and jungles became much more common subjects for landscape artists in the nineteenth century, with ambiguity, disorder, roughness, monotony and disharmony supplanting traditionally ordered and allegorical visions of nature.[17] Painters of nature began to leave the didacticism of religious orthodoxy and conservative moralism behind, and as they did, began to embrace swamps and other wild places. Swamps and jungles tend, with their profuse

and untamed vegetation, towards anarchy; in this sense, the mere representation of the swamp becomes a challenge to traditional moral values.[18]

A. R. Waud's 19th-century 'Sunset in the Mississippi Swamp'.

The works of A. R. Waud and Joseph Rusling Meecker exemplify this shift in swamp representation. Waud, whose work during the Civil War had reinforced his readership's conception of the swamps as menacing and forbidding, returned to the Florida swamps after the war with a renewed appreciation. Where before his focus had been on the hardships and terrors the swamps held for Union soldiers, he now described the swamps as a kind of sacred space, in which 'all the feathered and reptile brethren in their vast cathedral went on with their beads and paternosters reckless of the intrusion or possible heresy in their midst.'[19] Monstrous and sacred mingle as the chaotic swamps become a kind of natural church.

The artist Joseph Rusling Meecker painted stunning swamp landscapes throughout the 1870s and '80s. Having worked as a paymaster on a Union gunboat during the war, Meecker had observed the swamps at first hand, and his paintings, based on sketches he made during his service, capture a mysterious beauty in their interplay of shade and light.[20]

In the works of artists like Meecker, objective representations of nature give way to subjective; moral clarity to ambiguity. Moonlight, with its ambient, indirect glow, supplants penetrating, ordering sunlight in many swamp depictions of the era. Swamp paintings tended to be dominated by twilight landscapes, in-between states – neither one thing nor another. Swamp imagery, then, follows a general pattern of moving away from traditional, morally laden representations of nature into celebrations of ambiguity.

Other painters of the Romantic era, inspired by the age's new appreciation for the wild sublime, rendered swamps and marshes in myriad ways. Frederic Church, a painter of the Hudson River School of Romantic landscape artists, painted

Joseph Rusling Meeker, *The Land of Evangeline*, 1874, oil on canvas.

scenes that emphasized the awe-inspiring magnitude of nature. Drawn to jungle landscapes, Church brought his sensibility to tropical swamps in paintings like *Morning in the Tropics* (1858), a jungle scene in which people in a tiny boat are dwarfed, in typically Churchian style, by the immensity of the jungle around them. Another Hudson River School artist, Régis François

Harry Fenn, 'A Florida Swamp', illustration from William Cullen Bryant's *Picturesque America* (1874).

Gignoux, celebrates the beauty of the swamps in works like the landscape *View, Dismal Swamp, North Carolina* (1850). Gignoux bathes the gnarled and shadowy swamp in the red-golden light of sunset, presenting a vision of the swamp as idyllic and peaceful.

Martin Johnson Heade, a landscape painter deeply fascinated by wild, beautiful, potentially menacing landscapes, rendered a variety of wetlands in his work. Heade captured the salt marshes of his native New England with an eye towards exploring the play of light within the distinctive landscape. Travelling through the tropics, Heade also painted scenes of dense jungle vegetation, emphasizing light and shadow, and the tension between beauty and chaos in the tropical flora. Late in his career, Heade turned his attention to the Florida marshes. An avid hunter, Heade balanced a sportsman's adversarial relationship with nature with an aesthetic appreciation of the fascination and beauty of marsh, jungle and swamp landscapes.

Ironically, the rehabilitation of swamps in the popular imagination was in part driven by a consumer culture that was increasingly able to experience the swamps as tourist destinations, to be appreciated aesthetically for their exoticism. Tourists from other parts of America and Europe began to flock to exotic, intriguing Southern swamp locations in the decades following the American Civil War as part of a newly emerging consumerism that cast such places as objects and experiences for consumption.[21] People with money and leisure time wanted to see these dramatic, gloomy, picturesque places, and were willing to pay to do so. This consumerist shift is ironic in several ways. As we have seen, swamp-dwelling cultures have often been characterized by outsiders and historians as anti-consumerist, which typically led either to celebration of their simple lack of acquisitiveness or, more often, derision for their lack of industry. If consumer culture is the end of development, as the necessity for production is supplanted by the necessity for consumption to maintain a healthy economy, swamps would seem its natural enemy. Beginning in the nineteenth century, though, the desire to clear and develop the swamps was joined by the desire to

'consume' them through the comparatively (but not entirely) benign conquest of tourism.

In America, alongside this burgeoning consumerism was the draw, for many, of a purer, simpler place – as troublesome as that description is when applied to the antebellum South. The Confederacy's defeat in the Civil War prompted a late nineteenth-century Romantic revisioning of the South as a whole and of the swamps in particular. During the era of reconstruction, the South in the estimation of the nation as a whole followed a similar path to the one the swamps in general have done over the centuries: once it was eliminated as a practical threat, the plantation South was increasingly celebrated and romanticized as an alternative to a fast-modernizing, urbanizing and mechanizing culture. 'Lost Cause' literature like the novels and stories of Thomas Nelson Page presented a plantation South inhabited by august and kindly chivalric lords, elegant, exalted ladies and happy, content slaves. Here, such works suggested, was the last possible alternative to the bustle of modernity – a life lived in connection with, rather than in competition with, the land and nature. Tourists who came to the South intrigued by it as a unique, quaint and picturesque region gravitated to the swamps

Kilmer's Swamp Cure, evidence of the consumerist turn in swamp rehabilitation.

in increasing numbers, eager to see these wild landscapes that popular rhetoric had identified so strongly, for better or worse, with Southern culture.

By the mid-nineteenth century, travel writers too began to embrace the swamps in new ways, echoing Romantic evocations of mystery, magic and picturesque gloom. T. Addison Richards's *The Romance of the American Landscape* (1854) already shows characteristics of a trend that would emerge in earnest in the decades following the Civil War. Richards refers to 'the elfish beauties of the mystic swamps',[22] and presents the swamps' darkness and menace as sources of pleasure rather than of misery:

> 'My favorite haunts,' said Mr Blueblack, 'are the dark and poisonous lagunes which lead into the mysterious heart of the ghostly swamps. Creeping in my canoe through these dismal passages – their black waters filled with venomous snakes and lurking alligators, and shut out from the light of day by the intervening branches of

A congregation of egrets in water.

the cypress, the dark foliage of the magnolia, and the
inextricable veils of rampant vine, with the gay trailing
moss pendant everywhere in mournful festoons – my
fancy has run riot through a thousand wild and dreary
imaginings which it would harrow up your soul to hear!'[23]

The perspective here is quite different from Thoreau's, of course:
instead of positing swamps as pure spaces, Mr Blueblack finds
imaginative inspiration in their menace, poison, gloom and
melancholy. Increasingly free of practical danger and superstitious stigma, the swamps become a playground for the Romantic
imagination.

One of the most famous nineteenth-century travel books
was the two-volume *Picturesque America*, published in 1872 and
1874 and edited by the noted poet William Cullen Bryant.
Richly illustrated with wood and steel engravings, *Picturesque
America* became a tremendous commercial success, and fed the
national appetite to tour America's natural, unspoiled places.
Excerpts from *Picturesque America* reflect a transitional point in
swamp representations, as they combine appreciation for the
swamps' beauty with the thrilling aura of suspense and potential
menace:

Our little craft bumps along from one cypress-stump to
another, and fetches up against a *cypress-knee*, as it is termed
. . . glancing off, it runs into a roosting-place of innumerable
cranes, or scatters the wild-ducks and huge snakes over the
surface of the water. A clear patch of the sky is seen, and
the bright light of a summer evening is tossing the feathery
crowns of the old cypress-trees into a nimbus of glory, while
innumerable paroquets, alarmed at our intrusion, scream out
their fierce indignation, and then, flying away, flash upon our
admiring eyes their green and golden plumage.[24]

Glorious beauty is offset by darkness and danger, gaining force
through juxtaposition with inchoate, exotic menace. As the
travellers proceed through the darkening swamp, they become

A. R. Waud,
'Magnolia Swamp',
from William Cullen
Bryant's *Picturesque
America* (1874).

anxious about how their pilot will lead them safely through the
'Egyptian darkness'. Strange visions emerge from the gloom:

> No imagination can conceive the grotesque and weird forms
> which constantly force themselves on your notice as the
> light partially illuminates the limbs of wrecked or half-
> destroyed trees, which, covered with moss, or wrapped
> in decayed vegetation as a winding-sheet, seem huge
> unburied monsters, which, though dead, still throw about
> their arms in agony, and gaze through unmeaning eyes
> upon the intrusions of active, living men.

Beauty turns to darkness; darkness to visions of horror and death.
But rather than fear, fascination is the result. These are not lost
and desperate colonists seeing genuine peril in the swamps'

depths; rather, they are tourists, presenting imaginatively charged impressions for readers' vicarious pleasure. Another turn in the swamp journey brings out a final classic trope, as we move from monstrous horror to aestheticized desolation:

> Another run of a half-mile brings us into the cypress again, the firelight giving new ideas of the picturesque. The tall shafts, more than ever shrouded in the hanging moss, looked as if they had been draped in sad habiliments, while the wind sighed through the limbs; and when the sonorous sounds of the alligators were heard, groaning and complaining, the sad, dismal picture of desolation was complete.[25]

Picturesque America exemplifies the shift that occurs as swamps cease to be the indomitable, menacing spaces that they were through so much of history. Now more accessible, now considerably less threatening, they could be appreciated as spectacles, with all the old tropes that had made them so shunned and woeful now contributing to the thrill of visiting them, either as tourists or vicariously through art and travel writing.

Among the most distinguished travel writers of the era was the famous and celebrated poet Sidney Lanier, who embraced the Florida swamps as a kind of playground for the mind, a

Maurice Garland Fulton, 'The Lanier Oak', from *Southern Life in Southern Literature; Selections of Representative Prose and Poetry* (1917).

William McIlvaine, *The Chickahominy Swamp*, 1862, watercolour.

space that nourished the soul and fired the imagination. Lanier's book *Florida: Its Scenery, Climate, and History* (1875) stands out from the increasingly crowded field of post-Civil War Southern travel writing, a genre for which Lanier had little use. As he puts it in his introduction,

> it is not in clever newspaper paragraphs; it is not in chatty magazine papers; it is not in guide-books written while the cars are running, that the enormous phenomenon of Florida is to be disposed of . . . The question of Florida is a question of an indefinite enlargement of many people's pleasures and of many people's existences as against that universal killing ague of modern life.[26]

Lanier's idea of the swamp as pastoral antidote to modern bustle and ennui was not unique; it resonates with like-minded writers dating back to Thoreau and Bartram. Instead of passively contemplating their sultry beauty, though, Lanier saw in the swamps near limitless sparks to the imagination. His description of a passage through a Florida swamp leaps from one improbable association to the next:

> At first, like an unending procession of nuns disposed along the aisle of a church these vine-figures stand. But presently, as one journeys, this nun-imagery fades out of one's mind, and a thousand other fancies float with ever-new vine-shapes onto one's eyes. One sees repeated all the forms one has ever known, in grotesque juxtaposition. Look! Here is a great troop of girls, with arms reached over their heads, dancing down into the water; here are high velvet arm-chairs and lovely green fauteuils of divers pattern and of softest cushionment . . . Yonder is a bizarre congress – Una on her lion, Angelo's Moses, two elephants with howdahs, the Laocoon group, Arthur and Lancelot with great brands extended aloft in combat. Adam bent with love and grief leading Eve out of Paradise . . . It is a green dance of all things and times.[27]

Representing, perhaps, the ultimate end of the transformative process, Lanier renders a vision of a swamp freed from the over-determined ideas that had so long defined it, and rather than replacing those tropes with the Romantic and predictable, he allows his imagination to range over a wild and mercurial array of impressions.

Swamps, of course, have long been practical as well as conceptual playgrounds. Hunting and fishing have been part of the allure of wetlands since time immemorial. Louisiana's proud designation as 'the sportsman's paradise' comes largely from the hunting, fishing, boating and other recreational activities offered by its extensive wetlands. These traditions, strong in the South since colonial times and fundamental to the culture of most swampy areas, are fundamental to the quasi-mythic connection of Southerners to the land that grew out of the South's agrarian tradition and became, over the course of the twentieth century, increasingly associated with wilderness, bottomland and swamps once those areas lost their power and stigma. By the 1920s and

'30s in the United States, a general consciousness of wetlands as endangered spaces was growing. Ecologists were not the only ones paying attention. William Faulkner, perhaps the pre-eminent Southern writer of the twentieth century, combined a sense of cultural, ecological and spiritual loss in his complex but lyrical hymn to the vanishing wilderness, 'The Bear' (1942). Faulkner's story, built around the ritual of the hunt for Old Ben, a great bear who embodies the spirit of the wilderness, links the hunt to a ritual celebration of the dwindling Mississippi bottomlands, and leaves the reader with the sense that their passing leaves both humankind and the landscape immeasurably diminished. Literary-minded or not, sportsmen began to recognize the implications that swamp clearance and development held for their way of life.

Sportsmen and women, whatever their politics may be in other respects, are among the American South's most vocal environmentalists when it comes to wetland preservation. Some

Emile Jean Horace Vernet, *Hunting in the Pontine Marshes*, 1833, oil on canvas.

Road signs for swamp
tours, Thibodaux,
Louisiana.

of these hunters and fishermen, who stand alongside academics
and conservationists in organizations like the Georgia Wildlife
Federation, have proudly adopted the title 'Bubba Environmentalists'. The GWF, founded in 1936 by sportsmen, has since become
Georgia's pre-eminent conservationist organization; as they
proclaim on their website, 'Today, our members include bird
watchers, hunters, anglers, educators, gardeners, hikers – a
diverse group of individuals united by our concern and compas-
sion for the environment.' Ducks Unlimited, formed in 1937,
calls itself 'the world leader in wetlands and waterfowl conservation'.[28] Combining a focused agenda of advocating for habitat
preservation with an embrace and celebration of the culture and
tradition of hunting, Ducks Unlimited emphasizes the significance of wetland conservation to the survival of tradition and
recreation in the American South.

This model of linking preservation of the natural environment to culture and recreation plays out, in sometimes complex
and paradoxical ways, in wetland areas all over the world.
Almost universally threatened, to some extent or another, by
development, pollution and natural disaster, swamps must be
actively preserved if they are to survive, and such preservation
requires political will. Thus official government agencies as well

as private entrepreneurs offer, through swamp tours, easy and thorough access to spaces whose appeal lies largely in their pristine, untouched nature.

Contemporary swamp tours combine a wide array of traditional ideas and associations. Depending on the company, they might emphasize access to pristine nature, ghosts and horrors, cultural authenticity, extreme wildlife encounters or some combination of the above. Tours around the world target different audiences and emphasize different aspects of the wetlands. Tours of the Okavango Delta, for example, emphasize scarcity and exclusivity. The number of tourists allowed at any given time is strictly managed. Africa Travel Resource proudly proclaims that

Okavango Delta.

the Okavango Delta offers probably the most pristine, the highest quality and the highest priced safari in Africa ... Almost all of Okavango is divided into vast private concessions, where visitor numbers are strictly limited and the camps can generally only be accessed by light aircraft.

Okavango tours are quite expensive, in part because, as Africa Travel Resource's *Okavango Guidebook* explains, visitors pay to maintain the sense of pristine solitude. The Botswanan government has controlled the number of tourists allowed since the late 1980s, in what the *Guidebook* describes as 'a highly successful strategy of low volume, high value tourism'. The area around the delta is composed of a variety of private concessions, each of which is permitted only a very limited number of guests at a time. Artificial scarcity of tourists, then, becomes its own selling point:

> In the private concessions of Okavango you really can feel truly out in the wilds. On gamedrives you will only meet at most a couple of other vehicles from the same camp or its related neighbours. On walking and mokoro safari you and your guide should be all alone in the bush ... This truly is a magnificent isolation and one which really does seem pretty well priced when you realise quite what you are getting for your money.[29]

Exclusive access to a pristine wilderness, magnificent isolation – Okavango tours offer an upscale, expensive version of the Thoreauvian ideal.

Some Okavango tours emphasize other features – notably the marketing of traditional, 'primitive' culture. One, offered through the company Detour Africa, advertises 'Bushman Tribal Dancing' alongside wildlife viewing.[30] Others combine rugged adventure and tribal tradition with varying degrees of luxury: Audley Travel advertises tours in traditional mokoro canoes and accommodations ranging from tents along a trail to luxurious camps boasting lounges, bars and swimming pools.

Ecotours, which market not only solitude and pristine wilderness but environmental consciousness and benefit, have also become popular ways of experiencing the world's swamps. The intriguingly named Terra Incognita Ecotours offers a tour of the Pantanal whose small footprint is a major part of its appeal:

> For each participant a donation will be made towards the Conservation of the Pantanal ecosystem . . . This is an Ecotour that will make a difference to you, and to the areas we visit.
> We will employ local people.
> We will use locally owned and operated lodges and outfitters.
> We will use local goods and services.[31]

The page concludes with a warning, perhaps aimed at tourists accustomed to being accommodated by nature when they pay to tour it: 'We hope to encounter both wild Jaguars and Tapirs on this trip, however, sightings cannot be guaranteed.'

In America, perhaps predictably, tours tend to be marketed somewhat differently, though ecotours are becoming increasingly popular. Some tours emphasize ties to local culture, and tend to appeal to a different audience than the world travellers attracted to 'adventure travel' destinations like the Okavango and the Pantanal. Others play up the sense of excitement, featuring photos of gape-mouthed, razor-toothed alligators on their brochures and homepages. Swamp tour centres like south Louisiana and Florida offer tours for every taste and temperament, combining virtually all the traditional, contradictory elements of swamp tradition. Some emphasize Cajun culture, some ecology, some pristine wilderness, some creepy thrills, some extreme wildlife encounters. The tours offer to each customer the swamp that he or she seeks.

The Billie Swamp Safari, advertised through the Official Home Page of the Seminole Tribe of Florida, provides a succinct example of the contemporary multifaceted swamp tour. Promising 'a tour of 2200 acres of untamed Florida Everglades

preserved in its pristine state by the Seminole Tribe of Florida', A tour boat crossing
the Billie Swamp Safari tailors experiences to various visitor a Louisiana swamp.
tastes and interests. An Airboat Trail focuses on viewing animals
in their native, unspoiled environments, while a swamp buggy
ecotour combines up-close encounters with animals with a sense
of Seminole history and culture. For creature encounters with a
less ecological focus, Billie Swamp Safari offers both a Venomous
Snake Show and a Swamp Critter Show. No matter which version of the swamp tour visitors opt to take, they are all encouraged to visit the Swamp Water Café, which features 'alligator tail
nuggets, frog legs, or American fare for the faint of heart'.[32]

This same savvy entrepreneurial marketing of the contemporary swamps in a way that combines the appeal of nature,
cultural authenticity and wildlife circus can be found in earnest
in the Louisiana swamp tour industry. In his article 'Wilderness
Theatre: Environmental Tourism and Cajun Swamp Tours',
Eric Wiley provides a brief history of Louisiana swamp tours.
While Florida tourism had been popular since the turn of the
twentieth century, the first swamp tour company did not open

until 1979, when 'Alligator' Annie Miller, acting on an idea from the Terrebonne Parish Chamber of Commerce, opened one.[33] Dozens more have opened since, generally employing similar marketing while emphasizing various aspects of the swamp experience. Some, like Cajun Pride Swamp Tours, focus on a kind of cultural showmanship. Guides with names like 'Wild Man' and 'Rodeo Gator' become embodiments of the mythic persona of the fun-loving, happy-go-lucky swamp-dwelling Cajun. As Wiley explains, 'The Cajun guide has joined the ranks of other Cajun entertainers – stand-up comics, singers, story-tellers, preachers, and televised chefs – as a solo performer of Cajun culture, drawing on regional dialects, stories, and music in the creation of a persona.'[34] These guides are to Acadiana what Paul Hogan and the late Steve Irwin have been to Australia – exaggerated figures of ruggedness and fearless authenticity, distillations of the colourful, fictional Cajun swamp. As Wiley puts it, 'tourists are enticed not by swamps-as-swamps but by "Cajun swamps." "Alligator Annie," "The Cajun Man," and "Cajun Jack" are figures who interpret the wetlands through the medium of Cajun folk culture.'[35]

Cartoon Cajuns are only part of the draw, however. Swamp tours also market an uneasy balance of pristine, unspoiled nature with the promise of certain, plentiful and vivid wildlife encounters. Munson's Swamp Tours, for example, claims to be 'the most "authentic" swamp tour in the state of Louisiana bar none! . . . You will be in a wildlife environment that has not changed in hundreds of years, nature at it's [*sic*] most primitive level.'[36] For all the promise of unspoiled nature and authenticity, most swamp tours emphasize one thing over all others: wildlife encounters, specifically with alligators, the leviathan-like embodiments of swamp danger. Tour guides attract alligators, generally docile and neutral towards their presence, by scattering marsh-mallows in the water; the bold among them will sometimes coax a gator to jump out of the water and snatch a proffered treat from the side of the boat. Munson Swamp Tours' home-page features a guide's hands wrapped around an alligator's jaws, with the caption, 'JUST HOW CLOSE DO YOU WANT TO GET??'

Alongside the promises of pure, unspoiled nature, swamp tours also market access to wildlife that will approach and perform with the regularity of circus animals.

For those intrigued by the supernatural lore surrounding the swamps, some tours emphasize those elements, as well. Dr Paul Wagner's Honey Island Swamp Tour teases visitors with the legend of a Wookie-like beast, the Honey Island Swamp Monster, who is rumoured to inhabit the surrounding swamps. In New Orleans, Jean Lafitte Swamp Tours offers a Haunted Swamp tour on weekends in October that augments the swamps with special effects and local actors in costume as swamp beasts

Honey Island Swamp Tour near Slidell, Louisiana, in the 1970s.

Sign for Torres
Swamp Tours,
Louisiana.

and undead spooks. Local storytellers regale tourists with tales of various ill-fated folk who ventured into the swamps, never to return. As Norm Glindmeyer, the tour's producer, puts it, 'It's . . . more a show than anything else.'[37]

If contemporary swamp tours layer contradictory themes into a kind of postmodern quagmire of swamp representation, contemporary swamp art is a diverse field that encompasses exploration, elegy, political anger, and ecological activism. Contemporary artists who take the swamps as their subject are driven by a diverse array of motivations, and examining some of the ways that contemporary swamps are framed as and transformed into visual spectacles provides compelling perspective on the state of contemporary wetlands.

Greg Guirard, the artist whose quotation opens this chapter and whose photographs appear throughout this book, is a native of St Martinville, Louisiana, and has lived most of his life among the swamps of Louisiana's Atchafalaya Basin, balancing teaching, writing and photography with work as a boat driver, crawfisherman and farmer, among other things. Guirard captures images of the swamps themselves as well as of the people who fish, hunt and live among them. While his work is creative, striking and often surprising, Guirard presents his photographs

overleaf: Bald cypress at the lake's edge as the sun slips below the horizon.

181

in a strongly culturally and historically grounded context. He writes with Romantic fervour of the swamp-dwelling Cajuns and their connection to the landscape: 'there is an unexplainable, almost mystical connection between the Cajun fisherman and the swamp, a desire not only to see but to touch, to be part of the wildness itself, to live deliberately in concert with the big woods.'[38]

While Guirard's work celebrates the beauty and heritage of the Atchafalaya Basin, it cannot help but be heavy with the

Karen Glaser, *Within the Swamp, Roberts Lake Strand*, 2009.

knowledge of fragility and loss. Two of his best-known books, *Atchafalaya Autumn* volumes I and II, invoke the season as a metaphor for the endangered and evanescent state of the contemporary Louisiana wetlands. While the essays accompanying his photographs frequently offer practical suggestions and courses of action for saving the Atchafalaya Basin, the overall tone of his work is inevitably elegiac. As Guirard puts it in *Seasons of Light in the Atchafalaya Basin*:

> I'm wondering whether this place will last. When my youngest son is my age, will he be able to find beauty and solitude here? Will he want to? Will the Atchafalaya Basin become a waste dump, with waters too polluted to produce edible seafood? Will it become so crowded that seekers of wilderness solitude have to look elsewhere?[39]

Much of the poignancy in Guirard's work comes from the melancholy sense that capturing the swamps' beauty in these images cannot arrest their disappearance. The swamp, to Guirard, is history, culture, a region's distinctive way of life, and its loss is both cultural and intensely personal to the photographer. As he laments,

> For me it's like being terribly in love with someone who has changed and is changing in disturbing ways. Most of the beauty and spirit that I found so attractive in her years ago are still there, but some of her newer features are hard to deal with. If she continues to change, it is I who will be left alone, for I cannot bring myself to accept the transformation. I remember too well. I am too deeply rooted in the way we were.[40]

As Guirard captures glimpses of a fading culture and way of life, photographer Karen Glaser captures unconventional, often disorienting visions of the swamp that underscore the alienness of the land and waterscapes among us. Shooting swamps mainly underwater, Glaser presents 'unapologetically messy', ambiguous images that combine photographic genres of land-scape, documentary and street shooting, blending the frankly documentary with the ethereal and otherworldly.

Glaser's style fits the fundamental ambiguity of the swamps, as viewers of her photographs often find it difficult to deter-mine whether the perspective is from land or from underwater. Glaser never doctors or retouches her images, welcoming the dirt, debris, swamp water tannins and other elements that may cloud them, and calling such elements 'the living and breathing matter that seasons the soup and reflects, refracts, and bends the light to create its complexity'.[41] Like Sidney Lanier, Glaser sees in the ambiguity of swamp spaces endless creative possibility.

While she is eminently conscious of the dire ecological peril facing wetlands in America and all over the world, Glaser tends not to engage issues of ecology and culture directly in her work. Glaser is motivated, above all else, by the desire to 'show [people] what they haven't seen before', and she believes that, by creating 'seductive' images, she can *cause* people to think rather than [tell] them what to think'.[42] While the Louisiana native Greg Guirard captures images of a fading way of life, of a place so familiar that its transformation feels to him like losing a loved one, Glaser goes in search of the unusual, unfamiliar, unconventional and otherworldly underwater image. In this sense, Glaser brings to mind the explorers, like Thesiger and Middleton, who seek out the unfamiliar in the swamps and marshes for its own sake.

In recent years, two collaborative art installations have cap-tured two very different responses to the environmental and ecological perils facing world wetlands. On 20 April 2010, the Deepwater Horizon oil rig exploded in the Gulf of Mexico, killing eleven workers outright and beginning what would become the largest oil spill ever in American waters. Viewers of cable news programmes were presented with a constant under-water video feed of the gushing oil for more than a hundred days, and the spill's full impact on the gulf and the coastal marshes in Florida, Louisiana, and other states is impossible to calculate fully. To many, the spill was the most dramatic single episode in a long history of ecological irresponsibility, and a

Sunset in a swamp.

group of Louisiana artists responded in 2011 with an exhibit entitled 'Catalyst: Artists of Southern Louisiana Respond to the Gulf Coast Oil Crisis'. The exhibit included over eighty works by an array of artists, including Debbie Fleming Caffery, Allison Stewart, Robert Tannen and many others. The exhibit included photographs, paintings, sculpture, drawings and even video and performance pieces, and was united by a sense of outrage and urgency that informed all the pieces. It was intended to 'ignite an environmentally, politically and socially charged conversation' regarding 'the degradation of the Louisiana Wetlands, the crippling effects on the seafood industry and the historically insufficient regulation of the petroleum industry in the Gulf South'.[43]

The exhibit included subjects ranging from fishermen and oystermen to landscapes to reimagined classic works of art recontextualized for the Gulf disaster. In all, the works on

display expressed varying perspectives on grief, anger, concern and outrage, and ranged from openly political statements to humanist visions. Curator Robin Wallis Atkinson expressed hope that viewers would come away from the exhibit with

Cypress stumps and water hyacinth along the Sibon Canal, November 1985.

a more cohesive understanding of the state of affairs in Southern Louisiana and the Gulf Coast – in relation to

the BP disaster, to ongoing wetlands loss, to underregulated industrial practice and the fragility and interconnectedness between industry . . . and environment.[44]

Another kind of wetland art exhibit, equally driven by ecological concern but distinct in its methodology and aims, is the Cheng Long Wetlands International Environmental Art Project. Initiated in 2009 and continued on a yearly basis, the project selects artists from around the world to create outdoor sculpture installations in the village and marshes around the Cheng Long Wetlands site. Founded and curated by Jane Ingram Allen, the exhibit is part of an environmental education programme sponsored by the Kuan-Shu Educational Foundation, with the goal of raising awareness of the environmental plight of the area, one of the poorest in Taiwan. Climate change, human mismanagement and natural disasters have all contributed to the deterioration of the wetlands, as water levels rise and the land submerges, and the Cheng Long Wetlands Project uses works of public art to counteract the damage.

While it has had different themes over the years, the project always focuses on involving locals in the artistic process, and ensures that they can interact with the works on display. Artists work with local schoolchildren and villagers to help educate them about environmental stewardship, and produce works in collaboration with the local community; one year, the artists worked with local fishermen for a fishing-themed project.

Installations in the 2015 iteration are located in public areas of the village and in the wetlands themselves. The theme, 'Fragile: Handle with Care', emphasizes the delicate nature of the local wetland, and artists used recycled and natural materials from the wetlands nature preserve to create their works. Five artists were chosen from a pool of nearly 140 from 56 countries, and represent Australia, Canada, Italy, Japan, Taiwan, Switzerland and Germany (two artists represent multiple countries).

All of the works associated with the project are designed to be experienced and interacted with rather than simply regarded behind museum glass. One work, by Chao-chang Lee, a native of Yunlin County, is called *Heart of the Dragon*, a construct of oyster shells, bamboo and nylon ropes that fades in and out of view as the waters rise and recede, and can be physically entered by visitors. The project's ideal is to educate and inspire locals, especially children, so that eventually they will take a more active and intentional role in protecting and preserving the natural environment. It has already inspired imitators in Australia and Denmark.[45]

Since the mid-nineteenth century, as attitudes about nature and wild places began to modernize, swamps have undergone profound changes in the popular imagination. As tourist destinations and subjects for art, swamps are spaces for aesthetic contemplation as well as inspiration for environmental and political action. Even as traditional visions of swamps as dark, frightening places persist, redemptive, alarming and provocative representations continue to evolve and change with the landscape itself.

Postscript: Paradise Lost?
The Swamps' Uncertain Future

It is, perhaps, inevitable that the final chapter of any book about swamps will tend towards elegy. The story of the swamps, broadly told, is one of remarkable reversals. Swamps have gone from the pestilential quagmires of Columella's time to wetlands celebrated – and mourned – for their unspoiled purity. They have gone from being perceived as stubborn obstacles in the way of progress to its seemingly inevitable victims. And they have gone from havens for society's outcasts to spaces where fading, cherished cultures persist, however diminished by the forces of modernity. With each of these transformations, the old ways of seeing the swamps have not so much been supplanted as supplemented, rendering contemporary swamps landscapes defined by centuries of contradiction and paradox. The one unquestionable truth about contemporary wetlands is that they are almost universally threatened.

While there is a kind of tragic pastoral romance to the idea of swamps as fading remnants of a mysterious, forbidding but idyllic Eden, the very real threats to the world's wetlands are ubiquitous and sobering. Swamps that remain must be jealously guarded from all manner of threats, ranging from poachers preying on the animals of the Pantanal to industrialists to the oil industry. From Australia's Wingecarribee to the Florida Everglades, from the Okavango to the Great Dismal, wetlands the world over face mounting dangers, both human and otherwise.

While increasing awareness has led many world governments to take steps to protect wetlands, forces both natural and

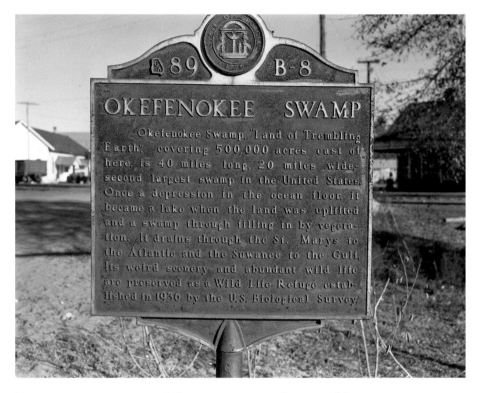

human continue to imperil these once seemingly invincible spaces. The marshes of southern Iraq have been largely destroyed, and with them, the culture of the Ma'dan. The Deepwater Horizon disaster and subsequent oil spill is only a particularly vivid one of many examples of human activity threatening the coastal marshes in the United States South; hurricanes like Katrina and Rita, and other storms that strike with increasing frequency, represent another danger. The many endangered species that dwell in wetlands around the world find their homes equally endangered since the age-old war between indomitable nature and human technological and industrial advancement has tipped so dramatically in the latter's favour.

A historical marker at the Okefenokee Swamp.

Assessments of the cumulative damage to the world's wetlands are dire, with some scholars estimating that around 90 per cent of the world's wetlands have been either damaged or destoyed.[1] In the United States, wetlands have been reduced by around half since the colonial era, from a total of around 89 million hectares to about 41 million hectares by the mid-1990s.[2] Even in places still largely regarded as unknown and impenetrable, like the swamps of Papua New Guinea, peoples who have lived for hundreds, even thousands of years virtually unchanged face threats to their homes and cultures from development, deforestation and climate change.

Fortunately, in the face of such dire threats, concerted efforts to slow, stop or even reverse wetland loss are underway around the world. The most powerful and wide-ranging of these efforts is that launched by the Ramsar Convention, a resolution adopted in and named for the city of Ramsar in Iran in 1971. The Convention, whose full title is the Convention on Wetlands of International Importance, is a treaty between world governments that 'provides the framework for the conservation and wise use of wetlands and their resources'. The oldest intergovernmental environmental agreement of its kind, the Ramsar Convention has been signed by close to 90 per cent of United Nations member states. The Convention defines wetlands broadly, including swamps, marshes, lakes, rivers, aquifers, peatlands and many other kinds, and has identified more than 2,000 sites for preservation in more than 160 countries. Focusing on fostering international cooperation to insure global sustainability, the Ramsar Convention's signees agree to 'work towards the wise use of all their wetlands', to 'designate suitable wetlands for the list of Wetlands of International Importance . . . and ensure their effective management' and to 'cooperate internationally on transboundary wetlands, shared wetland systems and shared species'.[3] This multinational effort is supplemented by efforts by a wide range of world governments, as well as private organizations, to preserve and sustain swamps, bogs and other wetlands. The headwinds they face are daunting: global climate change; pressures from industries like oil and timber; the understandable pressure, around the world, to spur economic growth at the wetlands' expense. In many places, like the marshes of southern Iraq, multinational efforts have met with some success in restoring formerly lost marshes. In many other areas, as in southern

Louisiana, where coastal marshes are disappearing at a rate of around 65 to 90 square kilometres per year, wetland preservation or restoration seems nearly impossible.

So what happens to swamps and swamp cultures in a world in which they are rapidly vanishing? Where once indomitable natural spaces are compromised, diminished, dying?

In his article on Cajun swamp tours, Eric Wiley describes a moment, late in the tour, when 'the wilderness fiction as a whole is on the wane'. The tours depend, Wiley explains, on a willing suspension of disbelief assailed, from the very beginning, by the realities of the contemporary swamp. Clear markers of the oil industry are everywhere:

> pipes, pumps, transport ships, warning signs posted along the canals, and the intricate, metallic structures that cap the natural gas wells, called 'Christmas trees' . . . the tourists' vision of the environment as a dangerous wilderness zone must overcome the pervasive actuality of an endangered zone in urgent need of protection.[4]

When the alligators summoned by tour guides bear scars from boat engine propellers, as often happens, the vulnerability of the swamp and its denizens is only underscored.[5]

Swamp peoples, as such, are imperilled by all the same forces that drive cultural change and evolution, exacerbated by the dwindling wetlands themselves. Seamus Heaney looks to the bogs of his homeland for preserved, buried cultural history in an imaginary act of cultural exhumation, but are all cultural relationships to swamp landscapes equally moribund? Even businesses that market the cultural aspects of the swamps tend to do so in a backwards-looking, nostalgic way. Okavango swamp tours feature dancing bushmen who perform for guests' pleasure; Seminole tours proudly revisit the swamp-dwelling past while the nation clearly focuses on a twenty-first-century entrepreneurial future. When I first researched the swamp tour industry in the early 2000s, I was struck by a passage, since removed, from the Cajun Pride Swamp Tours website that offered tourists the

The Atchafalaya Basin.

chance to 'view a trapper's cabin as your captain/guide teaches you about Cajuns – who they were, their origins, habits, lifestyle and other interesting historical facts'. The past tense puzzled me, particularly on a site whose very name asserted cultural pride: were the Cajuns dead, their culture consigned to history?

Marjorie Esman, in her article 'Tourism as Ethnic Preservation: The Cajuns of Louisiana', explains that contemporary Cajuns, having worked so hard to move beyond the practices of a culture long associated with deprivation and poverty and stigmatized by public schools that punished the speaking of French and encouraged assimilation, now function as 'tourist consumers of their own culture, visiting locales and events within their own

area for various purposes including the renewal of identity'.[6] Many keep hunting or fishing camps or bayou getaways, but few truly live among the swamps any more. Swamps, among other tourist destinations, 'recall the past and the traditions that have disappeared' to Cajuns. 'It is of little import that . . . nobody lives in the swamp anymore; these destinations suggest Louisiana exotica and they are promoted as ways to understand Cajuns.'[7] Ultimately, Esman argues, all of south Louisiana 'has become a tourist fiction, perpetuated by the tourist industry and by Cajuns themselves'.[8]

Cultural marketing at work: a sign for 'A Cajun Man's Swamp Cruise', c. 1997.

Esman may go too far, given the vital and significant role that the swamps still play in Acadian culture. It is difficult to deny, though, that those who still live among the Louisiana swamps more often than not make a conscious and considered choice to do so, to seek out ways of life consistent with their cultural heritage and fast-vanishing past. Cajun identity in the twenty-first century must be actively asserted and pursued.

However imperilled the actual swamps may be, however quaint and fading the idea of swamp cultures may seem, the very complex, compromised nature of the contemporary swamps continues to inspire provocative, powerful and significant art and literature. We have seen examples of swamps in contemporary horror films and art in previous chapters, but it seems appropriate to close with two examples of contemporary work that engage directly the problems and possibilities of the compromised and fading contemporary swamp.

The 2012 film *Beasts of the Southern Wild* is a particularly intriguing example of contemporary swamp representation. Widely celebrated, critically revered and yet at the heart of some controversy, *Beasts of the Southern Wild* presents a unique vision of a contemporary Louisiana swamp community that combines age-old tropes with ecological consciousness and a mythic vision akin to magical realism. The film centres on a young African American child named Hushpuppy, played by nine-year-old Quvenzhané Wallis, a native of Houma, Louisiana, whose performance made her the youngest recipient ever of a Best Actress Oscar nomination. Dwelling with her ill, alcoholic father and a small group of poor blacks and whites in the coastal community known as the Bathtub, Hushpuppy inhabits a man-made swamp, closed off from the mainland by a levee. Despite their poverty and isolation, Hushpuppy and her father Wink think of the Bathtub as a kind of paradise:

> Daddy says, up above the levee, on the dry side, they're afraid of the water like a bunch of babies. They built the wall that cuts us off. They think we all gonna drown down here. But we ain't going nowhere. The Bathtub's got more holidays than the whole rest of the world . . . Daddy always saying that up in the dry world, they got none of what we got. They only got holidays once a year. They got fish stuck in plastic wrappers. They got their babies stuck in carriers. And chicken on sticks and all that kind of stuff.

Hushpuppy's description draws on a veritable catalogue of swamp tropes: swamp-dwellers are separated from mainstream culture, in this case by a physical wall. They are free, in touch with nature. Implicitly anti-capitalist, they reject fish 'stuck in plastic wrappers' and 'chicken on sticks'. And, in a more positive echo of complaints about swamp people's industry dating back to colonial America, they have 'more holidays than the whole rest of the world'. Perhaps most important, they enjoy a connection with nature far beyond that of land-dwellers – Hushpuppy, who can hear the heartbeats of animals of all kinds and who enjoys an intuitive connection with even the great bestial aurochs who return from the melting polar ice in the wake of the great storm that threatens to destroy the Bathtub, exults, 'we's who the earth is for.'

The film is equal parts ecological commentary and primitivist, mythic fable. It displays both the global, large-scale effects of climate change and the intensely local victimization of the Bathtub, flooded by a storm that serves as a clear parallel to Hurricane Katrina until Wink bombs the levee to let the water out. It also draws explicit connections between the denizens of the Bathtub and the titular Beasts. Wink treats Hushpuppy with a kind of loving savagery, teaching her fierce strength and self-reliance while at times terrifying her with his drunken rages. Hushpuppy's moments of triumph come as she asserts her bestial nature, ripping the meat from a crab to the cheers of the gathered residents who had chanted for her to 'Beast it! Beast it!' The film's romanticized primitivism reaches its peak as Hushpuppy faces the huge, menacing, prehistoric aurochs and declares her kinship with them: 'You're my friend, kind of.' When the residents of the Bathtub are forcibly evacuated for their own safety, taken to a FEMA shelter that looks 'like a fish tank with no water', they mutiny and escape in a great stampede, returning to their fragile home in the flooded coastal swamplands.

Most viewers and critics were enchanted by the film: *New York Times* film critic Manohla Dargis praised it as 'hauntingly beautiful both visually and in the tenderness it shows toward the characters'.[9] While the bestial connections represent the people with a kind of primal, savage beauty, they risk falling into the essentialist traps that haunt so many popular representations of

Bird flying across
a swamp.

swamp people – and particularly people of colour. Writer and
critic bell hooks offers perhaps the most prominent critique of
a film that met with almost universal critical acclaim, reading it
as a Darwinian survivalist narrative with a subtext of violence and
degradation that the film's mythic overlay obscures or recontext-
ualizes: she calls it 'a ridiculous macabre fantasy of modern
primitivism'.[10] Whether we read the film as contemporary fable,
as flawed, primitivist exploitation, or as something in between,
it represents a postmodern vision of the contemporary swamp
as a combination of pure wildness and the work of man, a place
of magical beasts, exultant celebration, alternative communities
and intuitive connection with all living things, but also of disease, poverty, danger and the ever-present spectre of abandon-
ment. The combination of generations of swamp tropes with

timely (and timeless) ecological awareness and commentary renders a distinctly twenty-first-century swamp tale.

Sunrise at Henderson Lake, October 1993.

Another notable recent evocation of the swamp comes in Karen Russell's 2011 novel *Swamplandia!* While markedly different in tone and focus from *Beasts of the Southern Wild*, *Swamplandia!* is also a postmodern take on and revision of age-old swamp tropes, combined with an acute consciousness of the unique circumstances of the postmodern swamp. The novel plays with expected swamp iconography. It centres on a family of carnival performers who swim with and wrestle alligators in the titular park in the Florida swamps. The family represents, as we might expect, a connection to nature and to the environment, a kind of authenticity in the face of a monstrous swamp-themed amusement park, The World of Darkness, that opens up nearby. That authenticity is complicated by the fact that the family has long claimed a false Indian heritage as a selling point for its shows. Also, while the Bigtree clan has been raised in relative isolation out in the coastal Florida swamps, they are not anti-capitalist – the novel turns, after all, on the survival of their business, marketing swamp attractions for tourists. Like *Beasts of the Southern Wild*, the novel gives us a version of a man-made swamp, but here it is not levees and rising waters due to climate change that create it – it is created and marketed by competing interests, each of which sells a version of packaged tropes to outsiders.

Russell's swamps, like those in *Beasts of the Southern Wild*, are a combination of disparate threads and elements old and new. At times, they seem menacing and mysterious, full of quasi-supernatural threats, a portal to a stygian spirit-world into which Ava, the thirteen-year-old protagonist, must venture. Guided by the mysterious and sinister Bird Man, a Charon-like ferryman to another world, Ava tries to track down her sister, who has apparently absconded with her ghostly lover – a young

Sunset over a swamp.

Swamp sunset in
Cajun Country.

man who worked for the Civilian Conservation Corps in the
1930s, and who was killed in a boiler explosion on a dredge boat
– into the unexplored depths of the swamp. Though the mystical
overtones are eventually replaced by much more visceral and
earthly horror, the novel's swamps resonate with mystery and
mysticism.

At the same time, Russell's swamps are very clearly products of the post-industrial age. Choked with invasive, imported
melaleuca plants, strewn with remnants of past and present oil
exploration, they are presented for the consumption of tourists.
Magical, mysterious, ghostly, uncharted; invaded, threatened,
exploited, commodified – the swamp in Russell's novel combines nearly all its faces, factual and fictional. Russell presents
such a swamp in a novel that is itself a hybrid of genres, fusing
elements of Bildungsroman, Southern gothic, magical realism,
humour and horror.

Perhaps more closely and emphatically than any other natural feature, the swamps have reflected humankind's attitudes
towards nature itself, in their most polarized and powerful
forms. Even now, in an era when anyone with an Internet connection can see satellite photographs of virtually anywhere on
earth, the swamps retain their power to frighten us, to enchant

us, to inspire us, to move us. From foul fens and sickly marshes, to prelapsarian gardens and challenging frontiers, to endangered wetlands at last acknowledged for their significance to ecological balance and to human wellbeing, the swamps have transformed again and again over the ages, each time retaining some vestige of their prior identities in the popular imagination. They are, fittingly, sites of mixture, where reality and myth blend inextricably to imbue the landscape with unique and indelible meaning.

Appendix: An Array of Major World Wetlands

Wetlands can be found in every continent, with the exception of Antarctica. An encyclopaedic catalogue of world wetlands – even of 'major' ones – would be overwhelming.

The Ramsar list of Wetlands of International Importance, for example, includes more than 2,000 sites. Here, then, is merely a sampling of world wetlands of various types, selected based on size, variety and historical and ecological significance. I include it here to convey a sense of the many and varied landscapes and ecosystems we include under the umbrella term 'swamp'.

The Okavango Delta: Located in Botswana, the Okavango Delta varies in size due to seasonal flooding of many rivers, channels and other waterways, ranging from around 6,000 to over 20,000 square kilometres. The Okavango Delta serves as an enormous oasis, as animals travel through the vast Kalahari Desert to find food and water here. Vegetation in the Okavango Delta consists largely of papyrus, reeds and grasses. Wildlife in the Okavango Delta is largely itinerant, visiting seasonally for food and water before moving on. In addition to more than seventy species of fish, the wetland hosts elephants, giraffes, hippopotami, black and white rhinoceros, crocodiles, baboons and large numbers of Lechwe antelopes. During the high season, the Okavango Delta may serve as a temporary home for up to 200,000 large mammals, as well as more than four hundred bird species.

The Everglades: The Everglades, located in the southern part of the state of Florida, is one of North America's most famous 'swamps', though it does not technically qualify as one. A tropical, slow-flowing wetland sometimes called the 'River of Grass', the Everglades is more a region of sawgrass prairie than wooded swamp, though cypress swamps can be found throughout the larger wetland area, and mangroves are plentiful along the coasts. The wetland stretches around 100 km wide and more than 160 km long, and plays an important role in filtering water on its way to the Florida keys and coral reefs.

Facing rapid deterioration due to development and drainage, the Everglades have been the site of conservation, protection and restoration projects since the 1970s, and still face existential threats due to deteriorating water quality.

The Everglades are home to more than 350 bird species, many of which have been famously captured by the painter John James Audubon. While hunting and habitat loss have thinned the bird population since the nineteenth century, the Everglades National Park provides a sanctuary for them now. Some notable species include the white and glossy ibis, the roseate spoonbill, the wood stork, and the snowy egret. The Everglades are also home to land mammals including raccoons, white-tail deer and the famous Florida panther, as well as to reptiles including the American alligator.

The Sundarbans: Bangladesh's Sundarbans is a coastal mangrove forest, the largest of its kind in the world. The Sundarbans spans around 10,000 square kilometres, and encompasses much of southern Bangladesh as well as a small area in Eastern India. In addition to the saltwater mangrove forests, the Sundarbans region contains extensive freshwater swamps, which are home to a wide variety of flora and fauna, but which have experienced extensive habitat loss due to the development of the surrounding areas. The most famous species native to the Sundarbans is the royal Bengal tiger; the Sundarbans are home to what is likely the largest population of royal Bengals in the world.

The Mesopotamian Marshes: Also known as the Iraqi Marshes, the Mesopotamian Marshes have dwindled considerably since the mid-twentieth century, but were once among the world's largest wetland systems. The Mesopotamian Marshes are home to the Marsh Arabs, descendants of the ancient Sumerians. The marshes include the Central, Hammar and Hawizeh Marshes, and span around 3,000 square kilometres in southern Iraq as well as parts of Iran and Kuwait. The marshes' bird population has dwindled alarmingly since the efforts of Saddam Hussein in the early 1990s accelerated a draining process that has claimed more than 90 per cent of the total marsh area. Currently a variety of bird species, including the sacred ibis and the marbled teal, live in the Iraqi Marshes. The marshes are currently the focus of a multinational restoration effort.

The Atchafalaya Basin: The largest wetland area in the United States, south Louisiana's Atchafalaya Basin is an extensive forested wetland. Composed of cypress swamps and marshes, the Atchafalaya Basin spans around 5,700 square kilometres; motorists travelling between the state capital of Baton Rouge and the southwestern parts of the state can cross the Basin on the elevated Interstate 10. Culturally significant to the Acadian (or Cajun) people in particular, the Atchafalaya Basin has been a destination for hunters, fishermen and most recently ecotourists. The Basin is home to a wide variety of waterfowl and fish, as well as to the endangered Louisiana black bear. Like most of the wetlands on this list, it has suffered considerable degradation since the beginning of the twentieth century, much of it due to the region's oil and gas industry. The Basin now contains a National Wildlife Refuge.

The Pantanal: The largest wetland region in the world, South America's Pantanal spans somewhere between 150,000 and 200,000 square kilometres, and includes parts of Brazil, Paraguay and Bolivia. Composed primarily of floodplain, the Pantanal is largely submerged during the rainy season, which occasions a dramatic rise in water levels throughout the area. A tremendously biodiverse area, the Pantanal is home to around 1,000 species of birds, as well as to hundreds of species of fish, reptiles and mammals, including an extensive population of jaguars in addition to capybara and caiman. Though it is now partially protected, the Pantanal faces a variety of threats, largely due to development and its attendant deforestation and damage to water quality. Almost all of the Pantanal's land is privately owned, with the vast majority dedicated to cattle ranching.

The Great Dismal Swamp: Around 4,000 square kilometres in size, the Great Dismal Swamp is found in the southeastern United States. The swamp spans parts of Virginia and North Carolina, and includes a large lake, Lake Drummond, which has been memorialized in Thomas Moore's poem 'The Lake of the Dismal Swamp'. Forested with pine, cypress, cedar, oak and a variety of other trees, the swamp once spanned more than a million acres (400,000 ha), but has been considerably reduced over centuries of logging, development and deforestation. The 112,000-acre (45,000 ha) Great Dismal Swamp National Wildlife Refuge now serves as home to around two hundred species of birds, in addition to dozens of species of butterflies and a wide variety of reptiles and mammals, including black bears. The Dismal Swamp has great cultural significance as a haven for escaped slaves, whether passing through on the Underground Railroad or settling there and living in the swamp wilds as maroons.

The Sudd: A great South Sudanese swamp whose name translates as 'Barrier', the Sudd is a freshwater wetland and is part of the Nile basin. The Sudd's size fluctuates greatly depending on the season; generally, it is around 30,000 square kilometres, but can grow to over 120,000 in times of heavy rain. The Sudd is composed of a combination of grasslands and wooded areas, and, as it is situated among great expanses of desert, attracts hundreds of species of migratory birds and other animals, including hippopotami and crocodiles. As its name implies, the Sudd has become notorious as an obstacle to exploration and development, owing to the difficulty of navigating it as well as the large mosquito population that breeds in its waters.

The Vasyugan Mire: Siberia's Vasyugan Swamp, or Vasyugan Mire, is a vast freshwater swamp and peat bog. At 53,000 square kilometres, it is among the largest swamps in the world. It represents a remarkable ecological treasure. Uninhabited by humans, the swamp has massive peat reserves – more than one billion tons of peat, in fact. The Mire is suffused with clean, fresh water; dotted by lakes and crossed by a multitude of rivers, it is home to reindeer, elk, mink, otters and a variety of other animal species, as well as bird species such as the peregrine falcon and the golden eagle. The bog sustains a variety of endangered plant and animal species.

REFERENCES

Introduction: Terra Incognita

1 Ralph Tiner, *In Search of Swampland: A Wetland Sourcebook and Field Guide*, 2nd edn (New Brunswick, NJ, 2005), p. 3.
2 'The Great Vasyugan Mire', http://whc.unesco.org, 6 March 2007.
3 Ibid.
4 Ann Vileisis, *Discovering the Unknown Landscape: A History of America's Wetlands* (Washington, DC, 1997), p. 33.
5 Ibid., p. 73.
6 Charles M. Poser and George W. Bruyn, *An Illustrated History of Malaria* (New York, 1999), p. 20.
7 'The Great Dismal Swamp: A History', www.albemarle-nc.com, accessed 12 September 2016.

1 Swamp as Home: People of the Swamps

1 See Edward L. Ochsenschlager, *Iraq's Marsh Arabs in the Garden of Eden* (Philadelphia, PA, 2004).
2 Gavin Young, *Return to the Marshes: Life with the Marsh Arabs of Iraq* (London, 1977), p. 36.
3 J. M. Coles, 'Prologue: Wetland Worlds and the Past Preserved', in *Hidden Dimensions: The Cultural Significance of Wetland Archaeology*, ed. Kathryn Bernick (Vancouver, BC, 1998), pp. 3–26; pp. 8–9.
4 Cathryn Barr, 'Wetland Archaeological Sites in Aotearoa (New Zealand) Prehistory', in *Hidden Dimensions*, ed. Bernick, p. 53.
5 Bryony and John Coles, *People of the Wetlands: Bogs, Bodies and Lake-dwellers* (New York, 1989), p. 173.
6 Ibid.
7 Stuart McLean, 'Céide Fields: Natural Histories of a Buried Landscape', in *Landscape, Memory, and History: Anthropological*

Perspectives, ed. Pamela J. Stewart and Andrew Strathern (London, 2003), pp. 47–70; p. 50.

8 See Federico Borca, 'Towns and Marshes in the Ancient World', in Valerie M. Hope and Wrann Marshall, *Death and Disease in the Ancient City* (Florence, KY, 2000), pp. 74–84; pp. 75–6.

9 See ibid., p. 80.

10 Quoted in Coles and Coles, *People of the Wetlands*, p. 42.

11 Quoted ibid., p. 43.

12 Nick Middleton, *Surviving Extremes: Ice, Jungle, Sand and Swamp* (London, 2003), p. 196.

13 Ibid., p. 192.

14 Ibid., p. 232.

15 Ibid., p. 234.

16 Sam Kubba, ed., *Iraqi Marshlands and the Marsh Arabs: The Ma'dan, Their Culture and the Environment* (Reading, 2011).

17 Wilfred Thesiger, *Desert, Marsh and Mountain: The World of a Nomad* (London, 1979), p. 163.

18 Ibid., p. 168.

19 Young, *Return to the Marshes*, p. 33.

20 Ibid., p. 42.

21 Thesiger, *Desert, Marsh and Mountain*, p. 167.

22 Quoted in Young, *Return to the Marshes*, pp. 42–3.

23 See Kubba, *Iraqi Marshlands*, p. 12.

24 Thesiger, *Desert, Marsh and Mountain*, p. 167.

25 Edward L. Ochsenschlager, *Iraq's Marsh Arabs in the Garden of Eden* (Philadelphia, PA, 2004), p. 10.

26 Thesiger, *Desert, Marsh and Mountain*, p. 168.

27 Ibid., p. 170.

28 Ochsenschlager, *Iraq's Marsh Arabs*, pp. 11–12.

29 Stephen Adams, *The Best and Worst Country in the World: Perspectives on the Early Virginia Landscape* (Charlottesville, VA, 2001), p. 32.

30 Quoted ibid., p. 32.

31 Theda Perdue and Michael D. Green, *The Columbia Guide to American Indians of the Southeast* (New York, 2001), p. 41.

32 Michaela M. Adams, 'Savage Foes, Noble Warriors, and Frail Remnants: Florida Seminoles in the White Imagination, 1865–1934', *Florida Historical Quarterly*, LXXXVII/3 (Winter 2009), pp. 404–35; pp. 407–8.

33 Ibid., p. 412.

34 William Bartram, *Travels through North and South Carolina, Georgia, East and West Florida, the Cherokee Country, the Extensive Territories of the Muscogulges, or Creek Confederacy, and the Country of the Choctaws; Containing an Account of the Soil and Natural Productions of These Regions, Together with Observations on the*

Manners of the Indians, ed. Frances Harper (Athens, GA, 1998), pp. 15–16.

35 Ibid., p. 69.

36 Ibid., p. 36.

37 Clay MacCauley, *Fifth Annual Report of the Bureau of Ethnology to the Secretary of the Smithsonian Institution, 1883–84* (Washington, DC, 1887), pp. 469–532, at www.gutenberg.org.

38 Adams, 'Savage Foes', p. 417.

39 Ibid., p. 418.

40 Ibid., p. 426.

41 Charles W. Smith, 'Osceola's Seminoles Make their Last Stand', *New York Times Magazine* (26 April 1925), p. 14.

42 Quoted in Grunwald, *The Swamp*, p. 32.

43 Frances Kemble, *A Journal of a Residence on a Georgia Plantation*, in *Principles and Privilege: Two Women's Lives on a Georgia Plantation*, ed. Dana Nelson (Ann Arbor, MI, 1995), p. 76.

44 William Byrd II, *History of the Dividing Line: A Journey to the Land of Eden and Other Papers* (New York, 1928), p. 59.

45 Frank Shuffleton, introduction to *Notes on the State of Virginia* by Thomas Jefferson (New York, 1999), p. xxii.

46 Jack Kirby, *Poquosin: A Study of Rural Landscape and Society* (Chapel Hill, NC, 1995), p. 141.

47 Edwin T. Arnold, 'Introduction', in *Odd Leaves from the Life of a Louisiana Swamp Doctor* by Henry Clay Lewis (Baton Rouge, LA, 1997), p. xi.

48 Kirby, *Poquosin*, p. 161.

49 Ibid., p. 146.

50 Frederick Law Olmsted, *A Journey in the Seaboard Slave States in the Years 1853–1854, with Remarks on Their Economy* (New York, 1904), vol. II, pp. 332–3.

51 Harriet Jacobs, *Incidents in the Life of a Slave Girl* (New York, 1988), p. 113.

52 Olmsted, *A Journey in the Seaboard States*, vol. I, pp. 112–13.

53 Quoted in Kirby, *Poquosin*, p. 156.

54 Marjorie Esman, 'Tourism as Ethnic Preservation: The Cajuns of Louisiana', *Annals of Tourism Research*, XIII/1 (1986), pp. 451–67; p. 457.

55 Carl Brasseaux, *Acadian to Cajun: The Transformation of a People, 1803–1877* (Jackson, MS, 1992), p. 3.

56 Ibid., p. 4.

57 Ibid., p. 5.

58 Ibid., p. 21.

59 Ibid., p. 40.

60 Eric Wiley, 'Wilderness Theatre: Environmental Tourism and

Cajun Swamp Tours', *Drama Review*, xlvi/3 (Autumn 2002),
pp. 118–31; p. 125.

61 Ibid., p. 125.

62 Esman, 'Tourism as Ethnic Preservation', p. 453.

63 Ibid., p. 458.

64 Seamus Heaney, 'Feeling into Words', in *Finders Keepers: Selected
Prose, 1971–2001* (New York, 2002), pp. 15–27; p. 24.

65 Ibid., p. 24.

66 Seminole Tribe of Florida, 'Survival in the Swamp', www.semtribe.
com, accessed 21 October 2016.

67 Seminole Tribe of Florida, 'Seminole History', www.semtribe.com,
accessed 21 October 2016.

68 At www.semtribe.com.

69 Esman, 'Tourism as Ethnic Preservation', p. 457.

2 Swamp as Quagmire: Obstacle, Trial and Problem

1 Harriet Beecher Stowe, *Dred: A Tale of the Great Dismal Swamp*,
ed. Julie Newman (Halifax, 1992), p. 274.

2 Daniel Webster, 'Speech of Mr Webster at Capon Springs,
Virginia; together with those of Sir H. L. Bulwer & Wm. L.
Clarke, esq., June 28, 1851' (Washington, dc, 1851), pp. 7–8.

3 'The Swamp', *Chicago Tribune*, www.chicagotribune.com,
accessed 9 September 2016.

4 Ambrose Evans-Pritchard, 'u.s. asks nato for Help in "Draining
the Swamp" of Global Terrorism', www.telegraph.co.uk,
27 September 2001.

5 Columella, *De re rustica*, ed. H. B. Ash (New York, 1941), vol. i, p. 44.

6 Ibid., p. 42.

7 Charles M. Poser and George W. Bruyn, *An Illustrated History
of Malaria* (New York, 1999), pp. 21–2.

8 Jack Kirby, *Poquosin: A Study of Rural Landscape and Society*
(Chapel Hill, nc, 1995), p. 135.

9 Albert E. Cowdrey, *This Land, This South: An Environmental History*
(Lexington, ky, 1983), p. 37.

10 Susan Myra Kingsbury, ed., *The Records of the Virginia Company
of London*, 4 vols (Washington, dc, 1906–35), vol. ii, p. 374.

11 Bernard Romans, *Concise Natural History of East and West Florida*,
ed. Gail Busjahn, earlyfloridalit.net, accessed 3 September 2016.

12 Dianne Meredith, 'Hazards in the Bog: Real and Imagined',
Geographical Review, xcii/3 (July 2002), pp. 319–32; p. 320.

13 Ibid., p. 323.

14 Plutarch, *Life of Julius Caesar*, in *The Parallel Lives* (New York, 1919),
vol. vii, p. 579.

15 Kenneth A. Lockridge, *The Diary and Life of William Byrd II of Virginia, 1674–1744* (Chapel Hill, NC, 1987), p. 24.
16 Ibid., p. 6.
17 Ibid., p. 141.
18 William Byrd II, *History of the Dividing Line: A Journey to the Land of Eden and Other Papers* (New York, 1928), p. 53.
19 Ibid., p. 62.
20 Ibid., p. 74.
21 Charles Royster, *The Fabulous History of the Great Dismal Swamp Company* (New York, 1999), p. 82.
22 Ibid., p. 422.
23 Ulysses S. Grant, *Memoirs and Selected Letters* (New York, 1990), p. 299.
24 Thomas Mann, *Fighting with the Eighteenth Massachusetts: The Civil War Memoir of Thomas H. Mann*, ed. John J. Hennessy (Baton Rouge, LA, 2000), p. 68.
25 Bruce Stanley Wright, *Wildlife Sketches, Near and Far* (Dublin, 1962).
26 Michaela M. Adams, 'Savage Foes, Noble Warriors, and Frail Remnants: Florida Seminoles in the White Imagination, 1865–1934', *Florida Historical Quarterly*, LXXXVII/3 (Winter 2009), pp. 404–35; p. 422.
27 'The Florida Everglades', *New York Times*, 10 March 1889.
28 Gavin Young, *Return to the Marshes: Life with the Marsh Arabs of Iraq* (London, 1977), p. 9.
29 Ibid., p. 12.
30 Ibid.
31 J.R.R. Tolkien, *The Two Towers*, in *The Lord of the Rings* (Boston, MA, 1993), p. 652.
32 Ibid., p. 653.
33 Ibid., p. 654.
34 J. O. Wright, *Swamp and Overflowed Lands in the United States*, U.S. Department of Agriculture, Office of Experiment Stations (Washington, DC, 1907). Circular 76.
35 Edward King, *The Great South* (Baton Rouge, LA, 1972), p. 84.
36 Nelson Manfred Blake, *Land into Water, Water into Land: A History of Water Management in Florida* (Tallahassee, FL, 1980), p. 88.
37 Donald Hey and Nancy Philippi, *A Case for Wetland Restoration* (New York, 1999), p. 45.
38 Ibid., p. 2.
39 Ann Vileisis, *Discovering the Unknown Landscape: A History of America's Wetlands* (Washington, DC, 1997), p. 4.
40 Percy Viosca, 'Louisiana Wetlands and the Value of their Wildlife and Fishery Resource', *Ecology*, IX (1928), pp. 216–29; p. 217.

41 Quoted in Hey and Philippi, *Case for Wetland Restoration*, p. 47.
42 Frank Snowden, *Conquest of Malaria: Italy, 1900–1962* (New Haven, CT, 2006), p. 3.
43 Elisabeth Rosenthal, 'In Italy, a Redesign of Nature to Clean It', *New York Times*, 21 September 2008.
44 Federico Caprotti, *Mussolini's Cities* (New York, 2007), p. 207.
45 Snowden, *Conquest of Malaria*, p. 155.
46 Ibid., p. 159.
47 Sam Kubba, ed., *Iraqi Marshlands and the Marsh Arabs: The Ma'dan, Their Culture and the Environment* (Reading, 2011).
48 *Burning Bush: Saving Peat Swamp Forests in Indonesia*. Films on Demand. Films Media Group, 2009.
49 United States Geological Survey, *Louisiana Coastal Wetlands: A Resource at Risk*, USGS Fact Sheet, http://pubs.usgs.gov, accessed 2 September 2016.

3 Swamp as Horror: Monsters, Miasma and Menace

1 William Bradford Smith, 'Crossing the Boundaries of the Civic, the Natural, and the Supernatural in Medieval and Renaissance Europe', in *The Book of Nature and Humanity in Medieval and Early Modern Europe*, ed. David Hawkes and Richard G. Newhauser (Chicago, IL, 2013), pp. 133–56; p. 135.
2 *Beowulf*, trans. Charles William Eliot, in *Epic and Saga* (New York, 1910), p. 8.
3 Ibid., p. 40.
4 Smith, 'Crossing the Boundaries', p. 144.
5 Dante Alighieri, *The Divine Comedy*, vol. I: *Inferno*, trans. Mark Musa (Bloomington, IN, 1996), VII, pp. 106–26.
6 John Bunyan, *Pilgrim's Progress* (Oxford, 1967), p. 8.
7 Bryony and John Coles, *People of the Wetlands: Bogs, Bodies, and Lake-dwellers* (New York, 1989), p. 44.
8 Hans Christian Andersen, 'The Girl Who Stepped on Bread', in *The Complete Fairy Tales and Stories*, trans. Erik Hougaard (New York, 2011), pp. 607–8.
9 Ibid., p. 608.
10 Ibid.
11 Andersen, 'The Marsh-king's Daughter', in *The Complete Fairy Tales and Stories*, pp. 553–4.
12 Ibid., p. 554.
13 Coles and Coles, *People of the Wetlands*, p. 178.
14 Don Brothwell, *The Bog Man and the Archaeology of People* (Cambridge, MA, 1987), p. 11.
15 Ibid., p. 12.

16 Ibid., p. 32.
17 Ibid., p. 42.
18 Quoted ibid., p. 44.
19 Lars Larsson, 'Prehistoric Wetland Sites in Sweden', in *Hidden Dimensions: The Cultural Significance of Wetland Archaeology*, ed. Kathryn Bernick (Vancouver, BC, 1998), p. 71.
20 Stuart McClain, 'Céide Fields: Natural Histories of a Buried Landscape', in *Landscape, Memory and History: Anthropological Perspectives*, ed. Pamela J. Stewart and Andrew Strathern (London, 2003), p. 54.
21 Chris Moiser, 'Dragons of the Gambia', www.forteantimes.com, April 2006.
22 'Hunt for Gambia's Mythical Dragon', http://news.bbc.co.uk, 14 July 2006.
23 George Eberhart, *Mysterious Creatures: A Guide to Cryptozoology* (Santa Barbara, CA, 2002), p. 280.
24 Ibid., p. 279.
25 Ralph Izzard, *The Hunt for the Buru* (Fresno, CA, 2001), p. 45.
26 Eberhart, *Mysterious Creatures*, p. 77.
27 Oliver Ho, *Mysteries Unwrapped: Mutants and Monsters* (New York, 2008), pp. 53–4.
28 'Gwrach-y-rhibyn', www.celtnet.org, accessed 13 March 2014.
29 Wirt Sikes, *British Goblins: Welsh Folk Lore, Fairy Mythology, Legends and Traditions* (London, 1880), p. 216.
30 Ibid., p. 217.
31 'Gwrach-y-rhibyn', www.visitcaerphilly.com, 2009.
32 Tynes Cowan, 'The Slave in the Swamp: Effects of Uncultivated Regions on Plantation Life', in *Keep Your Head to the Sky: Interpreting African American Home Ground*, ed. Grey Gundaker (Charlottesville, VA, 1998), pp. 193–207; p. 204.
33 Camilla Jackson, interview, in *Born in Slavery: Slave Narratives from the Federal Writers Project, 1936–1938. American Memory*, Manuscript Division, Library of Congress, http://memory.loc.gov.
34 Sikes, *British Goblins*, p. 18.
35 Porte Crayon, 'The Dismal Swamp', *Harper's New Monthly Magazine*, September 1856, pp. 452–3.
36 John Pendleton Kennedy, *Swallow Barn: Or, A Sojourn in Old Dominion* (Baton Rouge, LA, 1986), p. 261.
37 Thomas Nelson Page, 'No Haid Pawn', in *In Ole Virginia, or 'Marse Chan' and Other Stories* (Nashville, TN, 1991), p. 163.
38 Ibid., p. 172.
39 Ibid., p. 186.
40 Ibid., p. 164.
41 Eberhart, *Mysterious Creatures*, p. 225.

42 Cajun Encounters Swamp Tours, 'Honey Island Swamp Monster: Fact or Fiction?', www.cajunencounters.com, accessed 6 August 2015.

43 Dana Holyfield-Evans, 'Honey Island Swamp Monster', www. swampmonster.weebly.com, 9 February 2012.

44 Robert Greenberger, 'Swamp Thing', in *The DC Comics Encyclopedia* (New York, 2006), p. 297.

4 Swamp as Spectacle: Paradise, Sanctuary, Aesthetic Object

1 J. M. Mendelsohn et al., *Okavango Delta: Floods of Life* (Windhoek, Namibia, 2010), p. 42.

2 Ibid., p. 14.

3 Ibid., p. 62.

4 Ibid., p. 53.

5 Frederick Swarts, 'Introduction', *The Pantanal of Brazil, Paraguay, and Bolivia: Selected Discourses on the World's Largest Remaining Wetland System* (Gouldsboro, PA, 2000), p. 4.

6 Chung Hwan Kwak, 'The Pantanal and the Pantaneiros: Heartfelt Challenges and New Opportunities', in *The Pantanal of Brazil, Paraguay, and Bolivia*, pp. 265–9; p. 265.

7 William Bradford, *Of Plymouth Plantation*, ed. Samuel Eliot Morison (New York, 1952), p. 84.

8 David C. Miller, *Dark Eden: The Swamp in Nineteenth-century American Culture* (New York, 1989), p. 48.

9 Ibid., p. 50.

10 Ralph Waldo Emerson, 'Nature', in *The Essential Writings of Ralph Waldo Emerson* (New York, 2000), pp. 3–43; p. 3.

11 Ibid., p. 6.

12 Henry David Thoreau, 'Walking', in *Walden, Civil Disobedience, and Other Writings* (New York, 2008), pp. 260–87; p. 262.

13 Ibid., p. 264.

14 Ibid., p. 275.

15 Ibid., p. 267.

16 Ibid., p. 276.

17 Miller, *Dark Eden*, p. 157.

18 Ibid., p. 161.

19 A. R. Waud, 'On the Mississippi', *Every Saturday Magazine*, 5 August 1871, p. 141. Quoted in Ann Vileisis, *Discovering the Unknown Landscape: A History of America's Wetlands* (Washington, DC, 1997), p. 105.

20 See ibid., p. 106.

21 Miller, *Dark Eden*, p. 9.

22 T. Addison Richards, *The Romance of the American Landscape* (New York, 2012), p. 102.

23 Ibid., p. 103.
24 William Cullen Bryant, ed., *Picturesque America* (New York, 1894),
    p. 134.
25 Ibid., pp. 134–6.
26 Sidney Lanier, *Florida, Its Scenery, Climate, and History. With
    an Account of Charleston, Savannah, Augusta, and Aiken; A Chapter
    for Consumptives; Various Papers on Fruit-culture; and a Complete
    Hand-book and Guide* [1875] (Gainesville, FL, 1973), p. 17.
27 Ibid., pp. 27–8.
28 'About Ducks Unlimited', www.ducks.org, accessed 12 September
    2016.
29 'Africa Travel Resource, Okavango Guide',
    www.africatravelresource.com, accessed 12 September 2016.
30 'Detour Africa, 8 Day Namibia to Victoria Falls Tour',
    www.detourafrica.co.za, accessed 12 September 2016.
31 'Terra Incognita Ecotours, Brazil', http://ecotours.com,
    accessed 12 September 2016.
32 'Billie Swamp Safari', www.billieswamp.com, accessed
    12 September 2016.
33 Eric Wiley, 'Wilderness Theatre: Environmental Tourism and
    Cajun Swamp Tours', *Drama Review*, XLVI/3 (Autumn 2002),
    pp. 118–31; p. 119.
34 Ibid., p. 120.
35 Ibid., p. 125.
36 'Munson Swamp Tours', www.munsonswamptours.com,
    accessed 3 July 2015.
37 Quoted in Laura McKnight, '6 Ways to Scare up Halloween Fun
    in New Orleans', *New Orleans Times Picayune*, www.nola.com,
    5 October 2011.
38 Greg Guirard, *Inherit the Atchafalaya* (Lafayette, LA, 2007), p. 167.
39 Greg Guirard, quoted ibid., p. 166.
40 Guirard, quoted ibid., p. 157.
41 Tom Hall, 'Karen Glaser Photographs', www.artswfl.com,
    accessed 12 September 2016.
42 Ibid.
43 CLA Press Release, quoted in Thomas B. Harrison, 'BP Oil Spill
    Inspired Artwork in "Catalyst" Exhibit', http://blog.al.com,
    11 July 2011.
44 Ibid.
45 C. A. Xuan Mai Ardia, 'Sustaining a Fragile Environment: Art
    in Taiwan's Cheng Long Wetlands', http://artradarjournal.com,
    29 May 2015.

Postscript: Paradise Lost? The Swamps' Uncertain Future

1 J. M. Coles, 'Prologue: Wetland Worlds and the Past Preserved', in *Hidden Dimensions: The Cultural Significance of Wetland Archaeology*, ed. Kathryn Bernick, pp. 3–26; p. 7.

2 James A. Schmid, 'Wetlands as Conserved Landscapes in the U.S.', in *Cultural Encounters with the Environment: Enduring and Evolving Geographic Themes*, ed. Alexander B. Murphy and Douglas L. Johnson, with the assistance of Viola Haarmann (Lanham, MD, 2000), pp. 133–56; p. 136.

3 'The Ramsar Convention and its Mission', www.ramsar.org, accessed 12 September 2016.

4 Eric Wiley, 'Wilderness Theatre: Environmental Tourism and Cajun Swamp Tours', *Drama Review*, XLVI/3 (Autumn 2002), pp. 118–31, p. 124.

5 Ibid., p. 123.

6 Marjorie Esman, 'Tourism as Ethnic Preservation: The Cajuns of Louisiana', *Annals of Tourism Research*, XIII/1 (1986), pp. 451–67; p. 454.

7 Ibid., p. 460.

8 Ibid., p. 464.

9 Manohla Dargis, 'Amazing Child, Typical Grown-ups', www.newyorktimes.com, 27 January 2012.

10 bell hooks, 'No Love in the Wild', http://newblackman.blogspot.com, 5 September 2012.

BIBLIOGRAPHY

'About Ducks Unlimited', www.ducks.org/about-du

Adams, Michaela M., 'Savage Foes, Noble Warriors, and Frail
  Remnants: Florida Seminoles in the White Imagination, 1865–
  1934', *Florida Historical Quarterly*, LXXXVII/3 (Winter 2009),
  pp. 404–35

Adams, Stephen, *The Best and Worst Country in the World: Perspectives on
  the Early Virginia Landscape* (Charlottesville, VA, 2001)

'Africa Travel Resource, Okavango Guide', www.africatravelresource.
  com/okavango-delta-botswana/guide

Alighieri, Dante, *The Divine Comedy*, vol. I: *Inferno*, trans. Mark Musa
  (Bloomington, IN, 1996)

Andersen, Hans Christian, 'The Girl who Stepped on Bread', in *The
  Complete Fairy Tales and Stories*, trans. Erik Hougaard (New York,
  2011), pp. 607–8

—, 'The Marsh-king's Daughter', in *The Complete Fairy Tales and Stories*,
  pp. 553–4

Arnold, Edwin T., 'Introduction', *Odd Leaves from the Life of a Louisiana
  Swamp Doctor* by Henry Clay Lewis (Baton Rouge, LA, 1997)

Barr, Cathryn, 'Wetland Archaeological Sites in Aotearoa (New Zealand)
  Prehistory', in *Hidden Dimensions: The Cultural Significance of Wetland Archaeology*, ed. Kathryn Bernick (Vancouver, 1998),
  pp. 3–26, pp. 47–55

Bartram, William, *Travels through North and South Carolina, Georgia,
  East and West Florida, the Cherokee Country, the Extensive Territor-
  ies of the Muscogulges, or Creek Confederacy, and the Country of the
  Choctaws; Containing an Account of the Soil and Natural Productions
  of These Regions, Together with Observations on the Manners of the
  Indians*, ed. Frances Harper (Athens, GA, 1998)

Beowulf, trans. Charles William Eliot, in *Epic and Saga* (New York, 1910)

Bernick, Kathryn, ed., *Hidden Dimensions: The Cultural Significance of
  Wetland Archaeology* (Vancouver, 2002)

'Billie Swamp Safari', www.billieswamp.com

Blake, Nelson Manfred, *Land into Water, Water into Land: A History of Water Management in Florida* (Tallahassee, FL, 1980)

Borca, Federico, 'Towns and Marshes in the Ancient World', in *Death and Disease in the Ancient City*, ed. Valerie M. Hope and Wrann Marshall (Florence, KY, 2000), pp. 74–84

'BP Oil Spill Inspired Artwork in "Catalyst" Exhibit', http://blog.al.com/entertainment-press-register/2011/07/bp_oil_spill_inspired_artwork.html

Bradford, William, *Of Plymouth Plantation*, ed. Samuel Eliot Morison (New York, 1952)

Brasseaux, Carl, *Acadian to Cajun: The Transformation of a People, 1803–1877* (Jacksonville, FL, 1992)

Brothwell, Don, *The Bog Man and the Archaeology of People* (Cambridge, 1987)

Bryant, William Cullen, ed., *Picturesque America* (New York, 1894)

Bunyan, John, *Pilgrim's Progress* (Oxford, 1967)

'Burning Bush: Saving Peat Swamp Forests in Indonesia', Films on Demand. Films Media Group, 2009.

Byrd, William, II, *History of the Dividing Line: A Journey to the Land of Eden and Other Papers* (New York, 1928)

Caprotti, Federico, *Mussolini's Cities* (New York, 2007)

'CLA Press Release', quoted in http://blog.al.com/entertainment-press-register/2011/07/bp_oil_spill_inspired_artwork.html

Coles, Bryony and John, *People of the Wetlands: Bogs, Bodies, and Lake-dwellers* (New York, 1989)

Coles, J. M., 'Prologue: Wetland Worlds and the Past Preserved', in *Hidden Dimensions: The Cultural Significance of Wetland Archaeology*, ed. Kathryn Bernick (Vancouver, 1998), pp. 3–26

Columella, *De re rustica*, ed. H. B. Ash (New York, 1941)

Cowan, Tynes, 'The Slave in the Swamp: Effects of Uncultivated Regions on Plantation Life', in *Keep Your Head to the Sky: Interpreting African American Home Ground*, ed. Grey Gundaker (Charlottesville, VA, 1998), pp. 193–207

Cowdrey, Albert E., *This Land, This South: An Environmental History* (Lexington, KY, 1983)

Dargis, Manohla, 'Amazing Child, Typical Grown-ups', *New York Times*, 27 January 2012, www.nytimes.com

'Detour Africa, 8 Day Namibia to Victoria Falls Tour', www.detourafrica.co.za

Eberhart, George, *Mysterious Creatures: A Guide to Cryptozoology* (Santa Barbara, CA, 2002)

Emerson, Ralph Waldo, 'Nature', in *The Essential Writings of Ralph Waldo Emerson* (New York, 2000), pp. 3–27

Esman, Marjorie, 'Tourism as Ethnic Preservation: The Cajuns of Louisiana', *Annals of Tourism Research*, XIII/1 (1986), pp. 451–67

Evans-Pritchard, Ambrose, 'US asks NATO for Help in "Draining the Swamp" of Global Terrorism', *The Telegraph*, 27 September 2001, www.telegraph.co.uk

'The Florida Everglades', *New York Times*, 10 March 1889

Grant, Ulysses S., *Memoirs and Selected Letters* (New York, 1990)

'The Great Dismal Swamp: A History', www.albemarle-nc.com/gates/ greatdismal

'The Great Vasyugan Mire', UNESCO World Heritage Centre, http://whc. unesco.org/en/tentativelists/5114

Greenberger, Robert, 'Swamp Thing', in *The DC Comics Encyclopedia* (New York, 2006), p. 297

Grunwald, Michael, *The Swamp: The Everglades, Florida, and the Politics of Paradise* (New York, 2006)

Guirard, Greg, *Inherit the Atchafalaya* (Lafayette, KY, 2007)

'Gwrach-y-rhibyn', Celtnet.org, www.celtnet.org.uk

'Gwrach-y-rhibyn', Cultural Caerphilly, 2009, www.visitcaerphilly.com

Heaney, Seamus, 'Feeling into Words', in *Finders Keepers: Selected Prose, 1971–2001* (New York, 2002), pp. 15–27

Hey, Donald, and Nancy Philippi, *A Case for Wetland Restoration* (New York, 1999)

Ho, Oliver, *Mysteries Unwrapped: Mutants and Monsters* (New York, 2008)

Holyfield-Evans, Dana, 'Honey Island Swamp Monster', www.swampmonster.weebly.com, 2 September 2012

'Honey Island Swamp Monster: Fact or Fiction?', Cajun Encounters Swamp Tours, www.cajunencounters.com/tag/letiche

hooks, bell, 'No Love in the Wild', Wednesday, 5 September 2012, NewBlackMan (In Exile), http://newblackman.blogspot.com/ 2012/09/bell-hooks-no-love-in-wild.html

'Hunt for Gambia's Mythical Dragon', BBC News, Friday, 14 July 2006, http://news.bbc.co.uk

Izzard, Ralph, *The Hunt for the Buru* (Fresno, CA, 2001)

Jackson, Camilla, interview, in *Born in Slavery: Slave Narratives from the Federal Writers Project, 1936–1938. American Memory*, Manuscript Division, Library of Congress, http://memory.loc.gov/ammem/ snhtml

Jacobs, Harriet, *Incidents in the Life of a Slave Girl* (New York, 1988)

'Karen Glaser Photographs', www.artswfl.com

Kemble, Frances, *A Journal of a Residence on a Georgia Plantation*, in *Principles and Privilege: Two Women's Lives on a Georgia Plantation*, ed. Dana Nelson (Ann Arbor, MI, 1995)

Kennedy, John Pendleton, *Swallow Barn: Or, A Sojourn in Old Dominion* (Baton Rouge, LA, 1986)

King, Edward, *The Great South* (Baton Rouge, LA, 1972)

Kingsbury, Susan Myra, ed., *The Records of the Virginia Company of London*, 4 vols (Washington, DC, 1906–35)

Kirby, Jack, *Poquosin: A Study of Rural Landscape and Society* (Chapel Hill, NC, 1995)

Kubba, Sam, ed., *Iraqi Marshlands and the Marsh Arabs: The Ma'dan, Their Culture and the Environment* (Reading, 2011)

Kwak, Chung Hwan, 'The Pantanal and the Pantaneiros: Heartfelt Challenges and New Opportunities', in *The Pantanal of Brazil, Paraguay, and Bolivia*, pp. 265–9

Lanier, Sidney, *Florida, Its Scenery, Climate, and History. With an Account of Charleston, Savannah, Augusta, and Aiken; A Chapter for Consumptives; Various Papers on Fruit-culture; and a Complete Hand-book and Guide, 1875* (Gainesville, FL, 1973)

Larsson, Lars, 'Prehistoric Wetland Sites in Sweden', in *Hidden Dimensions: The Cultural Significance of Wetland Archaeology*, ed. Kathryn Bernick (Vancouver, 1998), pp. 64–84

Lockridge, Kenneth A., *The Diary and Life of William Byrd II of Virginia, 1674–1744* (Chapel Hill, NC, 1987)

MacCauley, Clay, *Fifth Annual Report of the Bureau of Ethnology to the Secretary of the Smithsonian Institution, 1883–84* (Washington, DC, 1887), pp. 469–532, www.gutenberg.org

McKnight, Laura, '6 Ways to Scare up Halloween Fun in New Orleans', *New Orleans Times Picayune*, 5 October 2011, www.nola.com

McLean, Stuart, 'Céide Fields: Natural Histories of a Buried Landscape', in *Landscape, Memory, and History: Anthropological Perspectives*, ed. Pamela J. Stewart and Andrew Strathern (London, 2003), pp. 47–70

Mann, Thomas, *Fighting with the Eighteenth Massachusetts: The Civil War Memoir of Thomas H. Mann*, ed. John J. Hennessy (Baton Rouge, LA, 2000)

Mendelsohn, John, *Okavango Delta: Floods of Life* (Windhoek, Namibia, 2010)

Meredith, Dianne, 'Hazards in the Bog: Real and Imagined', *Geographical Review*, XCII/3 (July 2002), pp. 319–32

Middleton, Nick, *Surviving Extremes: Ice, Jungle, Sand and Swamp* (London, 2003)

Miller, David, *Dark Eden: The Swamp in Nineteenth-century American Culture* (New York, 1989)

Moiser, Chris, 'Dragons of the Gambia', *Fortean Times*, April 2006, www.forteantimes.com

'Munson Swamp Tours', www.munsonswamptours.com

Ochsenschlager, Edward L., *Iraq's Marsh Arabs in the Garden of Eden* (Philadelphia, PA, 2004)

Olmsted, Frederick Law, *A Journey in the Seaboard Slave States in the Years 1853–1854, with Remarks on Their Economy* (New York, 1904)

Page, Thomas Nelson, 'No Haid Pawn', in *In Ole Virginia, or 'Marse Chan' and Other Stories* (Nashville, TN, 1991), pp. 162–86

Perdue, Theda, and Michael D. Green, *The Columbia Guide to American Indians of the Southeast* (New York, 2001)

Plutarch, 'Life of Julius Caesar', in *The Parallel Lives* (New York, 1919)

Porte Crayon, 'The Dismal Swamp', *Harper's New Monthly Magazine*, September 1856, pp. 452–3

Poser, Charles M., and George W. Bruyn, *An Illustrated History of Malaria* (New York, 1999)

'The Ramsar Convention and its Mission', www.ramsar.org

Richards, T. Addison, *The Romance of the American Landscape* (New York, 2012)

Romans, Bernard, *Concise Natural History of East and West Florida*, ed. Gail Busjahn, http://earlyfloridalit.net

Rosenthal, Elisabeth, 'In Italy, a Redesign of Nature to Clean It', *New York Times*, 21 September 2008

Royster, Charles, *The Fabulous History of the Great Dismal Swamp Company* (New York, 1999)

Schmid, James A., 'Wetlands as Conserved Landscapes in the U.S.', in *Cultural Encounters with the Environment: Enduring and Evolving Geographic Themes*, ed. Alexander B. Murphy and Douglas L. Johnson, with the assistance of Viola Haarmann (Lanham, MD, 2000), pp. 133–56

'Seminoles Today', www.semtribe.com

Shuffleton, Frank, 'Introduction', *Notes on the State of Virginia* by Thomas Jefferson (New York, 1999)

Sikes, Wirt, *British Goblins: Welsh Folk Lore, Fairy Mythology, Legends and Traditions* (London, 1880)

Smith, Charles W., 'Osceola's Seminoles Make their Last Stand', *New York Times Magazine* (26 April 1925)

Smith, William Bradford, 'Crossing the Boundaries of the Civic, the Natural, and the Supernatural in Medieval and Renaissance Europe', in *The Book of Nature and Humanity in Medieval and Early Modern Europe*, ed. David Hawkes and Richard G. Newhauser (Turnhout, 2013), pp. 133–56

Snowden, Frank, *Conquest of Malaria: Italy, 1900–1962* (New Haven, CT, 2006)

Stowe, Harriet Beecher, *Dred: A Tale of the Great Dismal Swamp*, ed. Julie Newman (Halifax, 1992)

'Survival in the Swamp', www.semtribe.com

'Sustaining a Fragile Environment: Art in Taiwan's Chen Long Wetlands', http://artradarjournal.com, 29 May 2015

'The Swamp', *Chicago Tribune*, www.chicagotribune.com

Swarts, Frederick, 'Introduction', *The Pantanal of Brazil, Paraguay, and Bolivia: Selected Discourses on the World's Largest Remaining Wetland System* (Gouldsboro, PA, 2000), pp. 1–24

'Terra Incognita Ecotours, Brazil', http://ecotours.com

Thesiger, Wilfred, *Desert, Marsh and Mountain: The World of a Nomad* (London, 1979)

Thoreau, Henry David, 'Walking', in *Walden, Civil Disobedience, and Other Writings* (New York, 2008), pp. 262–78

Tiner, Ralph, *In Search of Swampland: A Wetland Sourcebook and Field Guide*, 2nd edn (New Brunswick, 2005)

Tolkien, J.R.R., *The Two Towers* (Boston, MA, 1993)

'United States Geological Survey. Louisiana Coastal Wetlands: A Resource at Risk', USGS Fact Sheet, http://pubs.usgs.gov/fs/la-wetlands

Van Young, Adrian, 'Santeria and Voodoo All Mashed Together: If You Want to Understand True Detective, You Have to Understand Louisiana', Slate.com, 4 March

Vileisis, Ann, *Discovering the Unknown Landscape: A History of America's Wetlands* (Washington, DC, 1997)

Viosca, Percy, 'Louisiana Wetlands and the Value of their Wildlife and Fishery Resource', *Ecology*, IX (1928), pp. 216–29

Waud, A. R., 'On the Mississippi', *Every Saturday Magazine*, 5 August 1871

Webster, Daniel, 'Speech of Mr Webster at Capon Springs, Virginia; together with those of Sir H. L. Bulwer & Wm. L. Clarke, esq., June 28, 1851', available at https://archive.org

Wiley, Eric, 'Wilderness Theatre: Environmental Tourism and Cajun Swamp Tours', *Drama Review*, XLVI/3 (Fall 2002), pp. 118–31

Wright, Bruce Stanley, *Wildlife Sketches, Near and Far* (Dublin, 1962)

Wright, J. O., *Swamp and Overflowed Lands in the United States. U.S. Department of Agriculture, Office of Experiment Stations* (Washington, DC, 1907)

Young, Gavin, *Return to the Marshes: Life with the Marsh Arabs of Iraq* (London, 1977)

ACKNOWLEDGEMENTS

A great many people have helped me on this journey through the swamps. I'd like to thank my friend and mentor Michael Kreyling for pointing me in this direction in the first place, many years ago. Thanks to the administration and my colleagues at LaGrange College for their support throughout this project. I also appreciate my fellow swamp scholars Eric Gary Anderson, Taylor Hagood and Kirstin Squint, whose work, insights and ideas continually reveal new depths to the swamp. Thanks also to Brannon Costello and to Bradford Smith for their advice and assistance.

A thousand thanks to my wife Jeanette and son Lucas for bearing with me and supporting me throughout this project, and to my mother, Professor Mary Ann Wilson, for her unflagging willingness to read and comment on chapters and drafts when she had so much else to do.

Finally, I'd like to thank the late Greg Guirard. Greg generously allowed me to use his magnificent photographs throughout this book. Moreover, he devoted much of his life to capturing, chronicling, defending and sustaining the Atchafalaya Basin. His writings and images balanced elegy and activism, celebration and melancholy, and always captured his infectious love for his native Louisiana and its landscape. He will be sorely missed, and I feel privileged beyond measure that his work appears here.

PHOTO ACKNOWLEDGEMENTS

The author and publishers wish to express their thanks to the below sources of illustrative material and/or permission to reproduce it. Locations of some artworks are also given below.

Vadim Adrinaov: p. 17; Alamy: pp. 38 (Anders Ryman), 73 (Gary Corbett), 82 (ART Collection), 96 top (© 20th Century Fox/Everett Collection), 96 bottom (Warner Brothers/AF Archive), 134 (United Artists GmbH), 139 (Moviestore Collection Ltd); www.freeimages.com: pp. 6 (Terri Heisele), 19, 21 (Hermo Sakk), 51 (elvis santana), 67 (Martin E. Jones), 110 (Thomas Pate), 158 (winterdorve); © Karen Glaser 2009: p. 184; Teo Gómez: p. 175; Greg Guirard: pp. 12, 13, 23, 63, 64, 95, 107, 122, 123, 152, 167, 182–3, 187, 188, 195, 199, 200, 202; Justin Hall: pp. 14–15, 144–5; iStockphoto: p. 164 (nicoolay); Library of Congress, Washington, DC: pp. 47, 80, 52, 55, 58, 80, 87, 88, 89, 90, 100, 155, 156, 171; LSU Libraries, Louisiana: pp. 174, 181, 196 (Dr Don Davis); Paul Mannix: p. 133; www.morguefile.com: pp. 9 (jppi), 22 (Paulabflat), 148, 178, 201 (photojock); Norbert Nagel: p. 125; National Archives and Records Administration (NARA): p. 54; National Library of Medicine: pp. 76, 102, 166; Jon Rawlinson: p. 146; Sven Rosborn: p. 119; Shutterstock: pp. 16 (neelsky), 24 (Roberto Tetsuo Okamura), 25 (elleon), 30–31 (Filipe Frazao), 32 (JFJfin), 35 (Eivaisla), 78 (Don Mammoser), 91 (Photocechcz), 147 (Filipe Frazao); State Archive of Florida: p. 192; State Library of Louisiana: pp. 68, 98, 106 (The Percy Viosca, Jr. Collection), 180; U.S. Fish and Wildlife Service: p. 81 (Laura Kennedy); Miguel Vieira: p. 18.

INDEX